Rave reviews for *The Science of Everyday Life*

"Ingram . . . acts as a kind of cross between a clear-eyed journalist and a tongue-in-cheek comedian." —*The Globe and Mail*

"Making science comprehensible and fun to read is no easy task. Ingram succeeds by making it seem easy."

—*The Edmonton Journal*

"Above all, Mr. Ingram encourages curiosity. And readers may find themselves pondering the nature of any number of everyday occurrences that Mr. Ingram never even got around to covering." —*The New York Times Book Review*

"*The Science of Everyday Life* is a delightful collection of essays to make you raise an eyebrow at the amazing 'worlds' which unfold as you go about your daily business." —*The Edmonton Sun*

"With such wide-ranging subject matter, there's bound to be something for almost every taste." —*The London Free Press*

"For those who like to ingest science in the form of intellectual hors d'oeuvres." —*The Washington Post*

"Especially welcome, and extremely rare . . . The quality and scope of Ingram's writing will lure many readers into scientific literature—an achievement that should not be underestimated."

—*The Ottawa Citizen*

"This is a great idea for a book and Ingram has realized the idea with wit and verve." —*Quill & Quire*

PENGUIN CANADA

THE SCIENCE OF EVERYDAY LIFE

JAY INGRAM is the author of the bestsellers *The Science of Everyday Life*, *The Burning House*, *Talk, Talk, Talk* and, most recently, *The Barmaid's Brain*. He is also the author of three children's books: *Real Live Science*, *A Kid's Guide to the Brain* and *Twins: An Amazing Investigation*. Currently the host of television's *Daily Planet*, Jay Ingram has written and hosted several radio documentaries, including "Cranial Pursuits" and "The Talk Show." *The Burning House* won the 1995 Canadian Science Writers Book Award. He writes a weekly science column in the *Toronto Star*.

Also by Jay Ingram

Talk, Talk, Talk

The Burning House

The Barmaid's Brain

The Velocity of Honey

Theatre of the Mind

For Children

Twins: An Amazing Investigation

Real Live Science

A Kid's Guide to the Brain

the SCIENCE *of*
EVERYDAY
LIFE

JAY INGRAM

PENGUIN
CANADA

PENGUIN CANADA

Published by the Penguin Group

Penguin Group (Canada), 90 Eglinton Avenue East, Suite 700, Toronto, Ontario, Canada M4P 2Y3
(a division of Pearson Canada Inc.)

Penguin Group (USA) Inc., 375 Hudson Street, New York, New York 10014, U.S.A.
Penguin Books Ltd, 80 Strand, London WC2R 0RL, England
Penguin Ireland, 25 St Stephen's Green, Dublin 2, Ireland (a division of Penguin Books Ltd)
Penguin Group (Australia), 250 Camberwell Road, Camberwell, Victoria 3124, Australia
(a division of Pearson Australia Group Pty Ltd)
Penguin Books India Pvt Ltd, 11 Community Centre, Panchsheel Park, New Delhi – 110 017, India
Penguin Group (NZ), cnr Airborne and Rosedale Roads, Albany, Auckland 1310, New Zealand
(a division of Pearson New Zealand Ltd)
Penguin Books (South Africa) (Pty) Ltd, 24 Sturdee Avenue, Rosebank, Johannesburg 2196, South Africa

Penguin Books Ltd, Registered Offices: 80 Strand, London WC2R 0RL, England

First published in a Viking Canada hardcover by Penguin Group (Canada),
a division of Pearson Canada Inc., 1989
Published in Penguin Canada paperback by Penguin Group (Canada),
a division of Pearson Canada Inc., 1990, 1994
Published in this revised edition, 2006

1 2 3 4 5 6 7 8 9 10 (WEB)

Original edition copyright © Jay Ingram, 1989
Revised edition copyright © Jay Ingram, 2006

Manufactured in Canada.

LIBRARY AND ARCHIVES CANADA CATALOGUING IN PUBLICATION

Ingram, Jay
The science of everyday life/Jay Ingram.

Includes bibliographical references.
ISBN-13: 978-0-14-305663-8
ISBN-10: 0-14-305663-8

1. Science—Miscellanea. I. Title.

Q173.I53 2006   500   C2006-902776-5

Visit the Penguin Group (Canada) website at **www.penguin.ca**

Special and corporate bulk purchase rates available; please see
www.penguin.ca/corporatesales or call 1-800-399-6858, ext. 477 or 474

To Cinzia

Contents

WELCOME TO THE
TONGUE-SHOW

Try to remember the last time you had to do something with great control and precision or concentration—like threading a needle or reading the instructions for your VCR. Was the tip of your tongue showing between your lips? If you're not aware of doing it yourself, think about the last time you saw a young child concentrating hard on connecting the numbered dots in a puzzle. I'd bet you saw the tip of his or her tongue then. Behavioural scientists call this peculiar little phenomenon "tongue-showing" (with their usual imaginative flair for naming), and it's a common and powerful form of non-verbal behaviour.

There are many ways of showing your tongue. You can just stick the tip of it out between your teeth and leave it there or flick it in and out or leave it folded over between your teeth without sticking it out or even point it to one side or the other. No matter what specific form tongue-showing takes, it's always done to convey the message, "Don't bother me." What's intriguing is that the message is sent and received unconsciously—neither the person showing his or her tongue nor the person at whom the display is aimed is aware of the intent or even the act itself. This immediately differentiates tongue-showing from sticking your tongue out at somebody or even running the tip of the tongue suggestively along the lips. These are both deliberate and conscious acts, and tongue-showing is neither. Yet it's a very common human signal, and it's possible to predict when someone will do it and what the effect will be on those that see it.

Children in nursery school tongue-show most often when they're involved in activities such as working with dough, drawing with crayons or even kicking a ball. The greater the concentration needed, the more often the tongue appears. In one experiment, as children made their way to the top of an ever-narrowing set of steps, their tongues started to creep out. Even children who are not tongue-showing to begin with start to do so the moment they catch an adult watching them.

The tongue-showing of pool-playing fraternity brothers at the University of Pennsylvania correlated directly with the difficulty of the shot, and the poorer players tongue-showed more than the good players. Even gorillas follow the pattern—they show their tongues more when engaged in demanding tasks like balancing on an upturned wagon or hanging from chains.

There are many other situations where tongue-showing has been observed, but the explanation is the same: people (or gorillas) show their tongues when they don't want to be disturbed by others, usually because they're involved in something that's demanding their total attention. It's a subtle form of social rejection, a signal to stay away from the tongue-showing individual. The setting may be unusual: the most dramatic tongue-show I have ever seen was produced by a woman riding a bicycle down the centre line of a busy downtown street in Toronto, her tongue fully extended. She was engrossed in keeping her bicycle steady, and her tongue-showing worked as far as I could tell: every car gave her as wide a berth as possible.

An experiment in Philadelphia illustrated how we rely on tongue-showing to keep others away. In this study a twenty-five-year-old white male sat at the top of a set of stairs leading into an office building. He stared intently at anyone who climbed the stairs to enter the building, but did not threaten actual physical contact. A hidden observer recorded the reactions of all those who approached the building alone. Their reactions were predictably varied: some looked away immediately, some smiled hesitatingly, others looked at their watches or brushed their hair.

But what was remarkable was the number that showed their tongues: seventeen out of a total of fifty. While most of these did it within a metre or so of the starer, some started tongue-showing more than three metres away, and surprisingly, three began only after they'd passed the starer. (Although this may seem a little late to send a non-verbal "leave-me-alone" signal, the experimenters reasoned that they had their signals ready just in case the starer pursued them.) Obviously the signal can be used to avoid social contact even when the tongue-shower is not busy. The controls in this experiment were fifty people who passed the same man sitting in the same place on the stairs, but the man was reading, not staring. Only two tongue-showed.

The proof that the tongue-showing message is received comes from an experiment involving twenty-five male and twenty-five female college students. A student was seated at a desk facing the back of a room and was told to complete a reading comprehension test. He or she then received a test booklet and was warned that every question had to be answered in sequence. The supervisor then said that he was going to be transcribing an audio tape and would wear headphones so as not to disturb the student. He sat at the front of the room, facing the student's back.

Page three had been removed from each booklet, and an observer hidden behind a two-way mirror noted each student's response as he or she made this unsettling discovery. Most students first swivelled in the chair and called out to the supervisor. He, of course, head bowed and ears covered with headphones—heavy, old-fashioned headphones—made no response. The student then had to either shout or walk up to the supervisor's desk to get his attention. The supervisor continued to concentrate on his work but showed the tip of his tongue to half of the students who were trying to interrupt him. The observer timed how long each student hesitated before calling out loudly or even tapping the supervisor on the shoulder, and because the observer couldn't see whether the supervisor was tongue-showing or not, the experiment was truly "blind."

The results provided dramatic evidence that tongue-showing really does deter others. Those students who saw no tongue-showing waited, on average, 7.72 seconds before interrupting the supervisor; students who were shown his tongue waited 19.93 seconds (and one student, apparently inordinately sensitive to this sort of signal, waited two minutes). This average difference of 12 seconds is not only statistically significant, it's two and a half times longer, which in the circumstances would have seemed like an eternity.

The most curious part of the study was the post-experiment interview: students were told the real purpose of the experiment and then asked to reproduce the facial expression of the supervisor. None depicted a tongue. Those who had actually seen tongue-showing by the supervisor denied they had seen it, although many of them admitted that they felt that the supervisor did not want to be disturbed. On the other hand, students who had *not* seen the supervisor tongue-showing reported feeling no reluctance to interrupt him.

So it seems that tongue-showing works, even though neither person involved is aware that the exposed tongue is sending a message. Why does the tip of the tongue convey the message, "Leave me alone"? Why not the nose or the eyebrows? One of the principal investigators in this area, Dr. Julia Chase of Barnard College, thinks it may have an early start. She observed a six-week-old infant lying in a crib with her tongue protruding. The child's mother explained that the baby did that when she was finished breastfeeding—too weak still to pull her head away, she pushed the nipple out with her tongue. The baby then left her tongue visible for some time, discouraging her mother from further attempts at nursing. Charles Darwin observed a six-month-old doing the same thing with novel foods. While these are only anecdotal accounts, they suggest that tongue-showing as a rejection signal might begin in infancy.

Once you are aware of it, you'll see tongue-showing everywhere, but note its variations: sometimes it will be disguised in situations where showing the tongue might be deemed inappropriate. An example cited by researchers is the televised beauty pageant, where finalists are asked challenging questions to demonstrate their range of knowledge. The tongue makes its appearance on cue, not peeking out between the teeth, but busily moistening lips that need no moistening.

THE ROADRUNNER
TRIUMPHS AGAIN

The Roadrunner and Wile E. Coyote are two of the many popular cartoon characters to have come out of the Warner Brothers studios in the period from the 1930s to the 1950s. Daffy Duck, Bugs Bunny, Elmer Fudd and Yosemite Sam are some of the others, but the Roadrunner and Coyote are unique in that physics plays a key role in their stories.

The Roadrunner/Coyote cartoons are, of course, just endless variations on the theme that the Coyote can never catch the Roadrunner, and in devising more and more complex schemes to trap the elusive bird, he succeeds only in injuring himself in more and more bizarre ways. Most of the time it's because physics is violated: things happen in the cartoons that cannot happen in the real world, and we laugh. The Coyote releases the catch to launch a boulder from a huge catapult, and the rock, rather than rocketing off into the air, barely clears the catapult and lands right on top of the Coyote. So the next time, he carefully positions himself on the other side of the catapult, out of the path of the stone. But when he releases the catch, the *catapult* flips over the stone and lands on the Coyote's head.

My favourite scenes are those in which the Coyote runs off the edge of a cliff (a gag that's used in many other cartoons as well). In one particular Roadrunner cartoon, the Coyote rents an Acme Super Outfit (his equipment always comes from the Acme company), speeds up to the edge of a cliff and leaps straight out over the chasm below, striking a Superman-like pose. He then slows, stops dead in mid-air, pauses, then plummets straight down like a stone. Sometimes the critical factor in these scenes is the Coyote's realization that he's hanging in mid-air—he's safe until he looks down and sees that he's three hundred storeys up. Only then does he fall.

We laugh because the idea that you could run straight off the edge of a cliff, then fall straight down is, well, laughable. Or is that really why we're laughing? Five percent of high-school and university students surveyed think that's exactly what happens if you run off a cliff. Five percent isn't much—that still means the vast majority realizes that the Coyote's straight-out then straight-down path *is* a cartoon view of the world. However, more than a third of that sensible majority of students (and so maybe half of the general population) opt for something more realistic: the Coyote would actually go straight out for a while, gradually start to curve down as his outward force diminished, then fall straight down as gravity asserts itself. Much more reasonable, but unfortunately just as wrong.

What path does the Coyote take over the cliff? A constant curve downward, a path that combines two motions in one. He is always moving away from the cliff, although ever slower because of air resistance, and he's accelerating downward because of gravity. But gravity doesn't have to wait for him to start to lose his horizontal speed—it starts acting the moment he's over the edge of the cliff. The Coyote moves out and down at the same time.

Not only do large numbers of people have no idea what would really happen if the Coyote ran off the edge of a cliff but, amazingly enough, many of them explain their wrong answers in exactly the same terms medieval philosophers would have used.

Straight off the cliff then straight down is just what the Arab philosopher Avicenna would have predicted in the eleventh century. The idea that the Coyote might go straight for just a little while, then curve gently and finally fall straight down is more modern—it didn't become really popular until the fourteenth century. The philosopher Albert of Saxony would have agreed with this description—it fitted very nicely a theory that he and others in the 1300s espoused, called the "impetus theory."

According to this theory, something that is flying through the air contains impetus put into it by whatever caused it to start flying in the first place. A spear contains impetus imparted to it by the movement of the spearthrower's arm. As the impetus gradually wears away, or leaks out, the spear slows down and gravity takes over. Wile E. Coyote runs off the cliff carrying with him an abundance of horizontal impetus, acquired by pushing his feet against the ground with every step. The impetus theory then predicts he'll continue horizontally for a while, followed by a gradually curving period as the impetus wears off and the effects of gravity begin to be felt. Then, finally, there's only gravity—the impetus has run out—so the Coyote falls straight down. It sounds pretty reasonable.

It took until the seventeenth century for first Galileo, then Newton, to show that the theory was completely wrong. There is nothing called "impetus" that enters the spear before it's thrown or the Coyote before he runs over the cliff. Newton turned the whole thing around in his first law of motion by saying that there doesn't have to be a force continually acting on moving objects. In fact it's the reverse. An object in motion will continue to move—forever—until an outside force (usually friction) slows it down. By putting the onus on stopping an object rather than on keeping it going, Newton changed the whole approach to the problem. But somehow three hundred years later, we still haven't got the message, and there are many other examples besides the Roadrunner and the Coyote to prove it.

Imagine there's a piece of plastic tubing lying on a pool table, the sort of tube that is attached to the back of your clothes dryer to carry the hot air outside the house. This particular tube is bent into a gradual curve, so that if you take a billiard ball and throw it in one end, it'll follow the curve of the tube and exit heading in a new direction. The question is, what sort of path across the table will the ball follow once it's left the tube?

There are several reasonable alternatives: the ball might continue to curve as it was inside the tube, or it'll curve for a while, then straighten out, or it might start to travel in a straight line the moment it leaves the tube. If you predicted any sort of curve, you were reverting to the fourteenth century. And again, this is a widely held belief—about 50 percent of students polled believe the ball will continue to curve. They apparently think the ball contains a supply of curved impetus. Newton would have known the right answer—once it's clear of the tube, the ball rolls straight, because there's no longer anything preventing it from doing so.

Take another example. You're given that same billiard ball to hold, and you're asked to walk rapidly across the floor to a spot where there's an X marked, and without stopping, let the ball go so that it hits the X as you walk over it. When do you let it go: before you've reached the X or just as your hand with the ball is directly over it or just after you've passed it? Of a group of high-school and college students surveyed, 50 percent thought you should let the ball go when you're right over the target.

The students' own accounts of what is happening reveal how they are still in the grip of the impetus theory. They explain that the ball would acquire some sort of horizontal motion if it were thrown, but not if it's passively carried. In their minds, a ball held in the hand acquires no impetus or velocity, and will drop straight down no matter how fast you're walking. Of course they are wrong. The ball is travelling at the same speed forward as you are, and you must let it go before you are over the X because it will continue to travel forward as it falls.

One of the curious things about this phenomenon is that children are actually better at coming up with the modern concepts when they're in kindergarten than they are later on. The combination of learning in the formal setting and becoming more familiar with the world around them somehow takes them away from modern physics and back to the Middle Ages— a little odd when you consider that children, of all people, have to know these things in their daily lives. Whether it's innocent activities, like running and bouncing a ball at the same time, or delinquent, like figuring out the ballistics of stones from a slingshot, knowledge of the movement of things is part of every kid's life. Yet while they're perfecting these actions daily, the under-standing of them is slipping away.

Even when these children have grown up and gone to college. Even when they've taken physics—not just one course, but several—there are still situations in which they make judgments based not on their sophisticated twenty-first-century physics knowledge but on impetus theory. One striking example was provided in a study of students at the University of California at Santa Barbara.

In this case the experimenters, Maria Kozhevnikov and Mary Hegarty, theorized that in most cases, we eventually learn, either by practice or by the book, that impetus theory doesn't describe the real world. But there will still be circumstances with which we are unfamiliar, and in those situations, we might just revert to the old impetus way of thinking. So they devised an experiment in which students watched a computer simulation of both a large and a small object rising straight up in the air. Unlike the Galilean experiment of dropping two different-sized balls from the Tower of Pisa, this is a rarely seen event. It also has the advantage of differentiating between the physics of Newton and impetus theory.

Impetus would predict that the larger object will rise more slowly than the smaller one because gravity will rob it of its impetus more quickly. Newton, on the other hand, would say that the air resistance felt by the smaller object would make it rise more slowly. So students watched as an object appeared to rise on the screen, then disappeared, and they were asked to remember the final resting position of the object. They then had to guess whether a new image that then flashed on the screen was in the same place, higher or lower than the final resting place of the rising object. If impetus theory had taken over, then they would have judged the smaller object to have reached a greater altitude than the larger—which is exactly what happened.

But what was most striking was that even expert physicists—in this case grad students and post-docs in physics—made the same impetus-based judgments. Kozhevnikov and Hegarty appeared to be right: when in doubt—even when you know better—try medieval physics!

What does this all mean? Well, obviously people can live their lives perfectly well without having the vaguest idea of how things move. It also proves that Galileo and Isaac Newton, two of the most outstanding scientists of all time, have made almost no impression at all on most people. Newton might be remembered for his apple, Galileo for dropping things from the Tower of Pisa, but that's about as far as it goes.

Let's not be depressed, though—there is a positive side to this. If you are one of those people who drop the ball directly over the target or think that you could run straight off a cliff, coast for a while and then drop straight down, you can at least take heart from the fact that you would have been right at home discussing concepts of motion with some of the most prominent philosophers of the Middle Ages. The ideas of Avicenna, Albert of Saxony and even Aristotle live on in your thoughts.

"When the Moon Hits Your Eye Like a Big-a Pizza Pie . . ."

Dean Martin had it right in his fifties hit: the moon does look like a gigantic pizza when it's just over the horizon. Sometimes it's so big you swear you can make out craters and plains on the surface that are not normally visible. As it climbs in the sky, it gets smaller, until when it's high in the sky, it's back to what most of us would call its "normal" size. But at the horizon, the moon seems obviously much bigger: experiments in which people were asked to match the size of the horizon moon to a set of discs projected on a screen revealed that they saw the horizon moon as being at least one and a half times bigger than the overhead moon, and sometimes even twice as big.

So why does the moon look bigger at the horizon? Many people assume it's closer. But it's not. The moon does vary a little in its distance from the earth, but that movement has nothing to do with its apparent girth when seen near the horizon. Nor is the huge horizon moon the result of some distorting effect of the atmosphere, although Aristotle, the first of a parade of scientists to struggle with this problem, thought that atmospheric moisture somehow altered the paths of light rays travelling from the horizon moon, making it appear to swell. He and everybody else who tried to explain why the horizon moon is bigger than the overhead moon were doomed to fail for one simple reason. It isn't bigger.

A simple test proves it: hold an Aspirin tablet at arm's length, and it will just cover the moon, no matter where it is in the sky. I know that's hard to believe, because in your mind's eye you can probably see the aspirin as a tiny disc against the giant whiteness of the horizon moon. But it works. Or you can simply take photos of the two moons, and see that the images are the same size. Measure them to be sure—these and more sophisticated determinations have been made, and they all conclude that the moon's image *is* the same size all the time. It looks bigger when it's near the horizon, but that is an illusion—it's all in our heads.

The man who should get the credit for realizing the key to the horizon moon problem is the eleventh-century Arab astronomer Alhazen, although for some strange reason his contribution has been largely ignored. Alhazen argued that it was a trick of our own perception that had nothing to do with the moon's orbit or the earth's atmosphere. And even though Alhazen's idea wasn't accepted for centuries, it's now taken for granted that we're somehow fooled into seeing the horizon moon as larger than when it's overhead—it's called the moon illusion.

Knowing it's an illusion doesn't explain how it works, and there are half a dozen theories—or more—that try. Forty years ago the favourite explanation was the "angle of regard theory," which claimed that having to lift your eyes to see the overhead moon causes you to judge its size differently. (Why this should be so was never clear, although some suggested it was caused by the movements your eyes have to make to focus on something above you.) However, if your eye angle were important, it should mean that if you were to lie down and look straight up at the overhead moon without craning your neck, it would look bigger, but it doesn't. While some scientists might agree there's a small effect caused by having to look up, most would add that it's by no means the whole answer to the moon illusion.

(One of the oddest theories was proposed in 1899, suggesting that when you're looking straight out at something on the horizon, gravity pulls on your eye and causes it to sag a little, increasing the distance between the lens and the retina, and so causing the image of the moon to get bigger. This theory, despite having its own peculiar charm, is now of historical value only.)

Alhazen said that the moon illusion is created because we think the horizon moon is farther away than the overhead moon, and this idea still plays a part in the most widely accepted explanation of the moon illusion, called the "apparent distance theory." When you look at the moon at the horizon, all the distance cues your mind employs suggest that the moon is far away, yet the image of that moon in your eye—the actual spot of light that falls on the retina—is the same size as when it's overhead, where there's nothing to indicate such great distance. This seems to be a paradox: the same object can't produce equal-sized images when it's both near and far, so your brain unconsciously decides that the horizon moon must be a larger object. Having made that leap, your brain goes one step further and makes the horizon moon actually look bigger. You are unaware that all this is happening in your brain—all you know is that the moon looks gigantic.

This explanation often seems strange at first glance, but there are some interesting bits of evidence that support it. First, it's clear that the horizon plays some important role: try looking at the horizon moon with your head between your legs. The illusion is suddenly greatly reduced, probably because you no longer have an underlying horizon to judge the distance to the moon. (Psychologists know that flipping a picture upside down reduces apparent distance.)

You can also arrange mirrors so that the overhead moon appears suspended over the horizon and looks big. Then do the reverse, projecting the larger horizon moon overhead, and it looks small. There doesn't have to be anything on the horizon or even anywhere on the terrain between you and the horizon for the effect to work; you're not judging the moon's distance by lining it up behind an endless procession of hills, trees and buildings. The moon illusion works perfectly over water or desert, where there's nothing to give direct cues as to distance or size.

This might be because we perceive the sky as if we're living in some gigantic domed stadium in which the roof is closer than the seats on the far side. It has been suggested that most people perceive the horizon as farther away than the zenith, the overhead point, even when looking at a perfectly blank sky. (We laugh at the idea of a flat earth, but we sometimes see the sky as an inverted soup bowl.) Place a moon in that sky and, of course, it would seem farther away when near the horizon. However, this idea of the sky as a flattened dome has fallen into disrepute, partly because the sky doesn't *always* appear to be flattened, and partly because objects like the moon seem to float in behind the so-called dome.

A psychologist named Zeno provided a beautiful example of the effect of the horizon on apparent size in the 1860s. If you have an after-image in your eye, like that left by a flashbulb, and you go outside and look at the overhead sky, you see the after-image superimposed on the blackness of the sky. Then switch and look down at the horizon, and the after-image looks bigger! Yet it can't be: it's just an overexposed area on the retina of your eye—it can't swell or shrink depending where you're looking. But somehow when it's superimposed on the sky near the horizon, it swells, suggesting that your brain is manipulating the apparent size of that after-image, just as it might with the moon.

You might have detected a paradox in the apparent distance idea—it argues that your brain creates the illusion of a bigger horizon moon unconsciously, the basis for this illusion being the greater apparent distance of the horizon moon vis-à-vis the overhead moon. The problem is, if you come right out and ask people how far away the bloated horizon moon is, they'll always say it's *closer* than when it's overhead. How can they think it's closer when their brains secretly think it's farther?

Well, psychologists who support this apparent distance explanation of the moon illusion are nothing if not resourceful. They start with the premise that the whole illusion goes on unconsciously, producing what appears to be a big fat horizon moon. Then, they speculate, our conscious minds take over, and (unaware of what the unconscious brain has done) say, "Hey, that moon is huge, it must be really close." Two steps, one unconscious, one conscious—it's the brain equivalent of the right hand not knowing what the left hand is doing.

It's these twists and turns in the apparent distance theory that have convinced critics it is not the final answer to the moon illusion. Although it is still probably the most widely accepted explanation, psychologists are still, millennia after the first theories, debating many different explanations. There are abstruse theories that demand a rethinking of large areas of perceptual psychology and theories that turn the problem around by suggesting it's not that the horizon moon looks so big, but that the overhead moon is abnormally *small*. There's even one theory that suggests the illusion is created because the moon is unlike any other object that moves overhead from the horizon. Overflying aircraft or cloudbanks seem to get bigger as they fly over us, but the moon doesn't. This theory claims that our brains make sense of this paradox by assuming the moon is actually moving *away* as it's rising, and so should appear biggest when it's right at the horizon.

Regardless of exactly why the moon illusion works, it is indisputably an illusion, and one of the most striking tricks of the mind you'll ever see. It's almost impossible to stand looking at that bloated horizon moon and convince yourself you're being deluded by your own brain. This is a perfect argument against the charge that science explains away nature, the idea that the world is somehow more charming if you don't know how it works. In this case, it's anything but. How much more fascinating to know that those storybook pictures of the Hallowe'en cat silhouetted against a giant moon are a manifestation of this trick your brain plays on you.

Sights and Sounds
in a Cup of Coffee

My friend Jearl Walker, a physicist and raconteur in the best Texas tradition, likes to cultivate an image of hard living: lots of beer after sunset, and lots of coffee as the sun rises. I'm indebted to him for the idea of looking into one of those endless cups of coffee and finding all kinds of interesting science.

One of the most striking examples is also one of the easiest: fill a cup to the brim with very hot black coffee and position it so light shines on the surface. A bright light bulb is fine, but early-morning sunshine is even better. You'll see a pattern on the surface of the coffee like flagstones in a path: irregular whitish patches less than an inch across, bordered by narrow dark lines. They'll shift back and forth, changing their outline from moment to moment, but they don't disappear. These whitish areas bordered by dark lines are called "convection cells," places where warm coffee is rising up from the bottom of the cup and cooler coffee from the surface is sinking back down.

That process, convection, is common in nature: any time warm air or liquid underlies cool, gravity seeks to pull down the cooler denser material, and the warmer rises to take its place. This happens on the earth, where the plates supporting the continents circulate down into the interior then back up, and even on the sun, where internal gases heated by nuclear fusion rise to the surface. This circulation is usually very orderly, and in laboratory experiments the ascending and descending streams of liquid can organize themselves into clusters of perfectly symmetrical columns, producing a honeycomb pattern at the surface of the liquid.

In your cup of coffee, the whitish areas are hot coffee coming to the surface and spreading out and the black lines are cooler coffee diving back down. Coffee is coffee—why should it be a different colour hot or cold? Actually it isn't. The whitish areas are, amazingly, droplets of water suspended in the air. When hot coffee reaches the surface, individual molecules of water from the coffee evaporate, tearing themselves away from the surface. Even though they are invisible, there are millions of them, enough to exert a powerful upward force as they fly into the air.

Once airborne, these fast-moving water molecules quickly cool, slow down and coalesce to form a layer of tiny but visible droplets of water, just above the surface of the coffee. In any other circumstances, the droplets would be pulled back down into the coffee by gravity, but in this case gravity is balanced by the buoyant force of the evaporating water molecules, so these droplets hang there, suspended less than a millimetre above the surface of the coffee. What you think at first glance are white areas on the surface of the coffee are actually low clouds hanging over the coffee. They're white because the droplets reflect light. The coffee underneath is black, and that's what you see at the edges of these convection cells where the cooler coffee is descending back into the cup. There's no updraft here, hence no suspended droplets, and so you have an unimpeded view of the surface of the liquid.

Sometimes, if you're lucky, you'll catch sight of a dark line cutting through the suspended droplets. It's there, then it's gone. This is a tiny whirlwind generated in the unstable hot air rising from the coffee—an unpredictable local weather disturbance. It collapses the droplet layer for a few thousandths of a second, then disappears.

Seeing these clouds of droplets requires a leisurely cup of coffee, but if you normally make your coffee in a rush, it's still possible to make scientific observations. For instance, before you add powdered coffee whitener to coffee (or even instant coffee to hot water), take a moment to clink a metal spoon against the side of your mug. You'll notice the sound is almost a musical note with a definite pitch.

But as you add the powdered coffee or whitener and stir the unappetizing mixture that results, the pitch of that clink changes. It can drop by an octave or more. Until 1969 no theories had been put forward to account for this phenomenon, one that must take place thousands of times every day in coffee rooms throughout North America. Finally a group of scientists at the Institute of Geophysics and Planetary Physics at the University of California at San Diego solved the mystery. After rejecting explanations like temperature effects, or the scattering of sound waves by particles floating in the brew, they suggested that air bubbles were the key.

Particles of instant coffee powder and whitener have, trapped on their surfaces, tiny bubbles of air. These are released when the powder dissolves in the water. Sound travels more than four times slower in air than water, so when the water fills with scores of bubbles, sounds in the mug will slow down. Even if the bubbles make up only one one-hundredth of the total volume of the coffee (a reasonable estimate for a spoonful of instant coffee), the speed of sound drops thirtyfold as compared with plain water.

The pitch of a note is actually a measure of the number of sound waves per second. So when the speed of sound drops in the coffee mug, there are fewer vibrations per second, and the pitch of the mug's clink drops, sometimes by as much as two octaves. As the bubbles clear out of the coffee by rising to the top of the mug, which may take up to a minute, the note rises too, eventually reaching its regular pitch.

Finally, if you buy take-out coffee on your way into work and you take cream in it, what should you do to ensure that your coffee will stay hot until you reach your desk? Add the cream right away, or wait and add it at the last minute before drinking?

In this case I can report my own results, because I have tried to answer this question more than once at my kitchen table. In a typical experiment I filled three identical coffee mugs with the same blend of coffee at temperatures between 75 and 75.5 degrees Celsius. (Although two would suffice, three gives you that extra confidence that your apparatus is working. Four or five cups would be even better, but the thermometers, cream, pen and paper for that many would be impossible for one person to manage.) I then added exactly one tablespoon of half-and-half cream to the first mug. The temperature dropped instantly to 68 degrees. A minute later I added cream to the second mug, then monitored the temperatures of all three mugs as they cooled. Although obviously the two mugs with cream were at all times cooler, the temperature gap between them and the remaining mug of black coffee closed as the minutes passed. After six minutes the black coffee was only 3 degrees warmer. When I finally added the cream to that mug, after seven minutes, its temperature dropped *below* the other two, and stayed that way.

This experiment, and others like it, allow me to draw two conclusions: if it takes you anywhere from four to seven minutes to start drinking your coffee after you fill your mug, you're better off to add the cream immediately—it'll be warmer when you drink it than if you add the cream at the last moment. The other is that coffee that has cooled to 54 degrees Celsius is about 10 degrees past its prime. Do anything you like with it, but don't drink it.

Initially the black coffee should cool faster than the white because there's a greater temperature difference between it and the room, but that seems not to be the whole explanation. It might be that the cream itself reduces the loss of heat from the coffee by convection, that process of hot coffee rising to the surface, cooling and dropping back again. Also, the addition of cream turns the coffee to beige and dots the surface with globules of cream. Both these differences may act to reduce the amount of heat radiating from the surface of the coffee.

So add the cream right away and enjoy the extra degree or two of warmth. Even better, add whipped cream, which insulates the surface of the coffee from the cooler air above it and keeps it hotter longer. Or, as Jearl Walker might suggest, try the experiment with Irish coffee, bearing in mind that repeatability is the essence of science.

SEX AND THE
SINGLE ARMREST

We all carry around with us an invisible bubble of personal space from which we usually exclude other people. Its dimensions vary, but it may extend 70 centimetres (a little over two feet) in front, 40 centimetres behind and 60 to each side. A wealth of social psychology research has established that we react negatively to violation of our personal space, because we reserve that area for ourselves. Obviously we can change the rules as we wish, allowing others into our personal spaces if we are fond of them, or if circumstances—such as a crowded subway—preclude maintaining the usual space.* Experiments have shown that people of high status have larger personal spaces (it must swell as your prominence grows), that reactions to violation of personal space can range from flight to violence, and that men maintain larger personal spaces than do women.

---

* Psychologist Robert Sommer has suggested that we cope with the loss of our personal space in subways by pretending that we are surrounded by inanimate objects, not people. Why else, he asks, would we be upset if someone apologizes for jostling us? By doing so, they force us to acknowledge that they are humans, destroying our little fantasy.

But while there is plenty of evidence that we cherish our personal space, it's difficult to design an experiment that delineates the extent of that space precisely. It is also tricky to quantify reactions to invasions of personal space. But one study brilliantly combined the two by asking the question: what happens when men and women put their personal space on the line, and fight for control of the armrests in the economy-class section of airplanes? This was the subject of a two-part study published by social psychologists Dorothy Hai, Zahid Khairullah and Nancy Coulmas, of St. Bonaventure University, in 1982. A total of 852 people on twenty different plane flights were watched to see who occupied the armrest, and then travellers were interviewed to get an idea of their attitudes toward possession of the armrest. Economy-class seats may well have been designed by an experimental psychologist, for their narrowness ensures direct person-to-person conflict over the armrest. You know if you've sat in such a seat that your arms naturally fall over the armrests, and there's *nothing* natural about the position you have to adopt if those armrests are denied to you. The armrest is actually an object that sits within the personal spaces of two individuals.

The 852 experimental subjects comprised 426 male-female pairs of passengers. (Sleepers and cuddling lovers were discarded.) The observations began after people had had a chance to settle in, eliminating the confounding effect of an early arrival grabbing the armrest, only to cede it later to a neighbour. Obviously size is a problem here too—a very large person in an economy-class seat may take possession of the armrests through physical necessity without even realizing it—so extreme size disparities between neighbours were also taken into account.

The results were clear-cut. Of the 426 man-woman pairs of passengers sitting next to each other, 284 men and only 57 women used the armrest: that's 5 men for every woman. In thirty-seven cases, both used it, and in a further forty-eight, the armrest remained untouched. Even when the data were further adjusted to ensure that only men and women of roughly equal size were considered, there were still three times as many men as women using the armrest.

A questionnaire distributed to about a hundred air travellers of both sexes illuminated this male in-flight aggression. Sixty-eight percent of males interviewed claimed it "bothered" them to have a seat-mate use the common armrest, while only 42 percent of the females felt that way. And it's males under forty who resent it the most. Of twenty-three who admitted it bothered them not to have the armrest, twenty-one said they felt "very annoyed" in that situation, especially when the transgressor was a female. "I feel I deserve to have it; she doesn't," was one comment; of three men who said they resort to verbal abuse in these situations, one added, "especially if it's a woman, since they don't need the space."

I confess to some feelings of possession about the common armrest myself, but only if I feel I've been the victim of some hyperaggressive armrest-grabbing boor who seems to assume that it's his ... or hers. I then spend the next ten minutes figuring out how I can take possession. Sometimes I manage it gradually, by resting just the tip of my elbow at the back and then slowly pushing it onto the main section, or even by sliding my forearm onto the extreme edge, then pushing gently sideways. Sometimes an all-at-once seizure is possible (especially easy if the possessor a gift of a gold crown, but he had some suspicions that the gold-smith had cheated him by adulterating it with silver. Such contamination wouldn't have been visible, and the problem was how to tell if indeed there was silver in the crown.

The answer is supposed to have come to Archimedes just as he was getting into the bath. As he did so, he realized that the volume of water that slopped out as he lowered himself into the tub must be equal to the volume of the part of his body that was submerged. That moment of insight supposedly inspired Archimedes to come up with the crucial test that would confirm or deny the crown's purity.

Archimedes knew that gold is more dense than silver, so a crown made of pure gold would have a smaller volume than a crown of equal weight containing both gold and silver. But since there was only one crown, the disputed one, what could he compare it with? That's where the bathtub came in: a kilo of gold dumped into a tub will displace the same volume of water whether it is a shapeless lump or a delicately worked crown. Archimedes only had to find a piece of gold of exactly the same weight as the crown and then compare the volume of water displaced by each. If there was silver in the crown, as King Hieron suspected, it would displace more water than a lump of pure gold of equal weight. It did, revealing the fraud of the crownmaker. (You didn't think he'd be innocent after all that?)

However, this ancient historian's account of simply immersing the lump of gold in a container of water, measuring the rise in the water level, then replacing the lump with the crown and measuring *that* change in the water level wouldn't have worked. As Archimedes expert Chris Rorres has pointed out, using a vessel that would be wide enough to admit the crown (which in turn had to fit over the king's head) would mean that the change in water level would be distributed over such a large surface area that any appreciable difference between pure gold and gold/silver would be indistinguishable from changes that would result from the inevitable errors that could creep in. Air bubbles clinging to the crown would yield a falsely high reading of its volume; drops of water lost as objects are lifted out of the water would have the opposite effect.

Rorres suggests that what Archimedes could have done was to place both objects on a two-pan scale, then submerge the entire scale. Because the silver/gold mixture has a greater volume, it displaces more water, and therefore experiences a great buoyant force, and so would appear to weigh less than the lump of gold. Rorres argues that this method would have made use of both Archimedes' law of buoyancy and his law of the lever (in the arm of the scale). For much more see **www.math.nyu.edu/~crorres/Archimedes/contents.html**.

Archimedes deserves much more than just recognition for introducing *eureka* into our vocabulary. He lived from about 287 to 212 BC, and has been called antiquity's most celebrated mathematician. The above puzzle is only a part of his book *On Floating Bodies*, and that in turn is only one of many works he wrote. He was a mathematician first, but legends abound of his ability to build machines to demonstrate the principles he knew to be true. He is supposed to have said to the same King Hieron that any weight, no matter how large, could be moved by any force, no matter how small. "Give me a place to stand on, and I will move the earth" is how it's phrased in the famous version, although I also like "If there were another world, and I could go to it, I could move this one." At any rate, Hieron challenged him to prove this claim, and according to Plutarch, this is how it went:

> [Archimedes] fixed upon a three-masted merchantman of the royal fleet, which had been dragged ashore by the great labours of many men, and after putting on board many passengers and the customary freight, he seated himself at a distance from her, and without any great effort, but quietly setting in motion a system of compound pulleys, drew her toward him smoothly and evenly, as though she were gliding through the water.

This expertise with the mechanics of motion also made Archimedes a wizard of war machines. When the Roman fleet was besieging Syracuse in 212 BC during the Second Punic War, Roman sailors were so frightened of Archimedes' catapults, grappling hooks and cranes that dropped weights on the decks of their ships that they would panic if so much as a stick appeared over the city walls. There's even a fascinating suggestion that he set Roman ships on fire by having his soldiers stand in rows on a set of bleachers, holding their polished shields in front of them. The shields formed a huge reflecting mirror that focused sunlight onto the wooden ships standing offshore. There's still a debate about whether this could have happened, but there is no dispute about the reality of the *eureka* moment. It is something that should be remembered every time an aching body is lowered into the bath—a fitting tribute to Archimedes' inventiveness and mathematical skill.

Start your bath with Archimedes, and finish it with the problem of which way the water swirls as it goes down the drain. It should go down the drain in opposite directions on either side of the equator—counter-clockwise in the northern hemisphere and clockwise in the southern—because of what's called the Coriolis effect, a sometimes difficult thing to visualize.

To understand the Coriolis effect you have to start with the idea that at the equator, the earth's rotation to the east is carrying you along at about 1,600 kilometres an hour, but if you're at the North Pole, your speed is zero. Anywhere in between is in between. (Toronto is moving east at about 1,100 kilometres per hour.)

The result is that travelling through the air in a straight line anywhere in the northern hemisphere takes you to the right of your target. It sounds peculiar, I know, but it happens. For example, consider flying from Ottawa to Toronto. Ottawa, because it is farther north, is moving east more slowly than Toronto. By the time you reach the latitude of Toronto, its greater speed would have carried it east of your line of travel. Flying back poses the same problem: when you take off from Toronto you're already moving with Toronto's very rapid west-to-east motion, and that carries you off to the east of Ottawa, again missing your target to the right. Even if you're travelling roughly east to west in the northern hemisphere, you'll be deflected to the right. When the Germans were shelling Paris with Big Bertha in World War I, the Coriolis effect pushed each shell a kilometre and a half to the right of the target—over a distance of only 122 kilometres.

The Coriolis force is everywhere: it tries to push your car to the right as you drive, and it's even been suggested that polar explorers tend to circle to the right near the North Pole, and to the left in the Antarctic. But it's hard to imagine that this could be the result of a Coriolis effect—every step in contact with the ice or snow corrects it, just as tires grip the road and prevent your car from drifting.

When the plug is pulled in the bathtub, the water starts to move toward the drain, but it's deflected to the right (in the northern hemisphere) in the time-honoured Coriolis way. So rather than plunging straight down the drain, it circles it, moving ever closer, in counter-clockwise circles.

However, there is one small technicality: although water will *tend* to swirl in the bathtub as it nears the drain, the effect is so incredibly slight that you have to go to extreme lengths to confirm that the effect is even there. Calculations show that a large wash basin at the latitude of Boston—where the first crucial experiment on this subject was done—is moving with the earth's rotation at about one millimetre per *minute*!

To prove that the effect does occur, experiments were done in the sixties by two groups of scientists: one in Cambridge, Massachusetts, at the Massachusetts Institute of Technology, and one at the University of Sydney, in Australia. Both groups used a tank six feet in diameter, filled to a depth of six inches with water. The drain was three-eighths of an inch across, set in the middle of the tank, flush with the bottom. The Australian group added some features not included in the Massachusetts experiment: they made their tub out of wood to minimize the disturbing effects that heat from metal tub walls might have on the water, and they also put their tub in a small, cement-walled basement room with no windows, and allowed no one except the experimenters into the room while the experiment was running.

Even though both the American and the Australian scientists used six-foot tubs, not wash basins, they were forced to add tiny wooden floats to indicate the direction of swirling. (The Sydney group, true Australians that they were, used slices of wine cork.) In addition, virtually any unexpected event—a gust of wind, a heavy footfall, even pulling the plug—can ruin the experiment. That's why in both experiments the plug was pulled out from below, leaving the stilled surface of the water undisturbed. That's also why the two groups allowed the water to sit for eighteen to twenty-four hours after filling, the tub covered the entire time, before pulling the plug. They even took care when filling the tub to swirl the water in the opposite direction in which they expected it to flow out, so that if the water did drain out according to theory, it couldn't be attributed to leftover momentum from the filling.

These 1960s experiments confirmed what everyone had expected—the water does flow down the drain in different directions in the northern and southern hemispheres, and presumably we can chalk that up as yet another proof that the earth is spinning. (Any failure to get the experiment to work would have been blamed on the equipment or the procedure, not on the earth.)

Sadly enough, an effect that small is too easily swamped in your own bathtub by splashing and kicking, rubber duckies, errant sponges, even by the very act of pulling out the plug. If you do see the water swirl out the way it should, it's probably just luck. On the other hand, it's just possible that one day all the other influences will neatly cancel each other out, and the water will be swirling counter-clockwise (or clockwise, depending where you are) solely because the earth is turning. That's what I prefer to think when I see the water draining the right way.

No Redeeming
Features

The birds that share our cities are often objects of our scorn and annoyance, not admiration. The house sparrow is a prime example. Sparrows are too common to be adored, but I think our disdain is also prompted by the fact that sparrows remind us of the least admirable aspects of ourselves. We want our birds to be colourful, graceful woodland inhabitants, selflessly devoted to their mates, not urban, tunelessly loud, undistinguished in appearance, singlemindedly dedicated to food and sex, and not even exotic enough to migrate south for the winter. A closer, scientific look at the house sparrow proves my point.

Right from the beginning, house sparrows got off on the wrong foot in North America. They were imported in the 1850s, in the naive belief that they would combat inchworm and cankerworm infestations in the eastern United States. It's the same old story as the introduction of exotic species anywhere: a few pairs were brought in, and in less than a decade people were scrambling for ways to kill them off. The battle became particularly vicious in Boston, where ornithologists fought in public over the sparrow.

Thomas Mayo Brewer, a friend of Audubon, whose name lives on in the Brewer's blackbird, loved them: "They will ere long become one of our most common and familiar favorites." Well, two out of three ain't bad, as they say—they are common and familiar. His opponent was Dr. Elliot Coues, author of *Key to North American Birds*, a scientist who could be both charming and viciously disagreeable, who demanded scientific proof in all arguments yet believed in ghosts, and who turned some neat phrases in describing house sparrows as "sturdy little foreign vulgarians" and "animated manure machines," finally dismissing them as being "without a redeeming quality." Sparrow haters claimed that not only did the rapidly multiplying birds fail to kill off appreciable numbers of caterpillars, they also displaced more attractive native birds and damaged sunflowers, grapes and apples. These birds were unwelcome intruders.

It took until 1931 for the house sparrow to be admitted to the American Ornithologists' Union listing as an "American" bird, but the bad feelings lingered long after that. P. A. Taverner, in his *Birds of Canada*, written in 1945, couldn't even confine his scorn to the birds themselves: "The nests are great bulky untidy masses of straw and grasses and the tendency of these birds to fill down-spouts and load with litter every projecting architectural feature of buildings makes them objectionable." Taverner ended his section on the sparrow by listing traps, poison and systematic destruction of the nests as the most effective means of control. And this is from a bird lover?

Well, you say, all the foregoing is simply the unenlightened attitudes of the late nineteenth and early twentieth centuries. Maybe so, but even the objective approach of the ornithologist of the

1980s failed to make the house sparrow an endearing bird. In fact, that research revealed these as birds who spend almost all their time eating, and the rest of the time indulging in indiscriminate sex and shirking parental responsibilities.

Finding food is a central concern for all birds, consuming some 90 percent of their time, but sparrows do it in a particularly inelegant way, becoming, en masse, a huge eating machine. As city dwellers well know, you almost never see sparrows by themselves. They move in flocks organized to sweep the landscape for food, and a close look at these flocks reveals that they are not a co-operative effort but a temporary liaison between totally selfish birds.

From an individual sparrow's point of view, eating in a flock could actually be a bad thing, because many mouths are competing for the available food. On the other hand, it's safer to feed with others, because predators are usually less successful when attacking a flock, presumably because they're confused by a multitude of targets flying this way and that. Also, if an individual sparrow is only one of many, the chances of *becoming* dinner are obviously less. What's more important to a house sparrow, safety or food?

The answer is safety, as long as it doesn't interfere with food. It takes a sparrow anywhere from one-half to three-quarters of a second to "head-jerk": stop pecking, take a quick glance for attacking hawks, then lower the head again for more pecking. But the bigger the flock, the more birds to share the lookout duties, and while the time for each look stays the same, an individual bird can afford to take more time between looks. As long as there are at least some birds scanning the skies at all times, the rest are safe to eat as much as they can.

But obviously a large flock consumes the available food quickly, so each sparrow must also recognize when it's time to move on. A study at Oxford University showed that sparrows know when to leave a feeding flock by measuring the time between pecks. Seriously! When there's still lots of food left, every peck, no matter how fast, will yield a seed. But as the food runs out, the time between pecks becomes greater, and when that time (the "inter-peck interval") reaches three seconds and persists over three consecutive pecks, it's GUT: that's "Giving Up Time" in sparrow-researcher talk. (You'll note that the last reason sparrows form flocks is for the camaraderie—a notion that was put forward for flocking in general by biologist J. K. Cannon in the 1930s.)

But what about the well-documented observation that sparrows who happen upon a new food supply will chirp noisily, thereby attracting other sparrows to share in the find? Isn't this wonderfully altruistic? Not a chance. Careful analysis of this behaviour by Mark Elgar, then working at Cambridge University, showed that sparrows behave differently depending on whether the food is in one large piece or in crumbs. If the food is in crumbs, and therefore dispersed and presumably difficult for one bird to gather himself, the discoverer chirps loudly, attracting others who pounce on the food. But if the food is "non-divisible," for example, a single slice of bread, the first bird on the scene may just keep quiet and fly off with the whole slice.

The strategy is clear: if the food is chopped up into crumbs, a single bird would have to spend a long time trying to consume it himself, and that's risky; better to bring in others who, while they'll eat a lot of it themselves, will also share the job of looking out for danger, giving the discoverer a chance to eat some of it. But if the food is one piece, and you can get away with it in one swoop, then go for it!

It would be unfair to characterize house sparrows as being interested in food only. There is one other side to their life, and it was described nicely by a California ornithologist, Leon Dawson. After watching a pair of sparrows copulate fourteen times in a row with five-second breaks between, he could only describe what he saw as "the male suffering from satyriasis and the female from nymphomania." This was more anti-sparrow propaganda from the early days, but again, current research has revealed that the sex and reproductive lives of the house sparrow are, well, indiscriminate.

Female house sparrows indulge in what scientists call "conspecific brood parasitism"—they lay some of their eggs in other females' nests. Some birds victimize other species this way, but female house sparrows foist their eggs on members of their own species, their neighbours. It's not yet clear how often they do it, or just how disadvantageous it is to the female who's the unwilling and usually unknowing recipient, although from the evolutionary point of view it's a great waste spending time and energy raising someone else's offspring. (The point is to pass *your* genes on to the next generation, not someone else's.) There's some evidence that occasionally a female may be aware that there's an egg in her nest that she didn't lay, but apparently she doesn't remove that egg or abandon her nest. Some scientists have speculated that she adopts a tit-for-tat strategy, and lays her next egg in a neighbour's nest, thus opening the door to a domino effect: each bird whose nest is violated lays her next egg in another nest, and so on. Apparently if you're a female house sparrow it pays to be both sneaky and suspicious—keep intruders out of your nest, but fill everybody else's with your eggs.

Meanwhile the male house sparrows are mating with anyone who'll have them. This had been suspected for some time, because researchers had often watched as groups of males displayed in front of a single female, with the occasional male mounting her from behind as she watched the others bobbing up and down and fluffing their feathers in front. Now sophisticated DNA finger-printing techniques have made it possible for the first time to determine exact paternity in house sparrow offspring, and the news isn't good for those who love fidelity in their birds. One pair of sparrows, who turned out to be mother and *son*, produced three offspring from two broods. And while it was clear from the genetic data that two were actually fathered by the female's mate, her son (follow this so far?), the third seemed quite different genetically, and in fact had to have been fathered by another bird. So, DNA kit in hand, the scientists searched the neighbouring nest boxes for the male in question, and soon found him—next door.

That's the life of the house sparrow. A gritty, urban, all-too-recognizable way of life, if you ask me. To be fair, other birds share some aspects of the sparrow's lifestyle, but only house sparrows put it all together in such an unattractive package. And lest you think I was biased from the beginning in this matter, I gave the sparrows every opportunity to win me over. For two consecutive days I sat on my back porch researching and writing this chapter, while no more than five metres away a flock of house sparrows decimated a feeder full of sunflower seeds. Surely, I thought, there must be more to these birds than head-jerks, egg dumping and GUTs. I waited in vain. Finally the endless "cheep cheep cheep" (or is it "cheap cheap cheap"?) got to me. On the third day, I didn't refill the feeder.

However, as a reminder that even the most successful species should always be looking over their shoulders, the life of the house sparrow has become much more precarious since I first wrote this chapter. Astonishingly, their numbers in England have been plummeting over the last few years, and it's clear that many local populations will simply disappear. When that begins to happen, the remaining populations, now without the possibility of being replenished by immigration, may soon bite the dust too.

It seems that changes in farming practices might be responsible. Most house sparrows live on or near small farms, where they can pick up enough stray grain to stay well fed, especially during the winter, when natural food sources are hard to come by. These days, with British farmers sowing some of their crops at different times of year, and keeping their grain in bird-proof containers, the sparrows are challenged. On one farm, provisioning the sparrows one winter temporarily halted the slide of a population that had declined by 80 percent over the previous thirty years.

It's hard to know what will happen to urban populations. It is true in eastern North American cities that I have seen house finches (whose populations are on the rise) aggressively displace sparrows from feeders in the same way the sparrows do to other birds, but whether that would be enough to seriously damage the house sparrow population isn't clear. There are some indications that the sparrows' numbers in cities have been in slow decline since the mid-1980s, but there is no definitive trend yet.

The Dynamics of the Cocktail Party

Cocktail parties are events of great interest to scientists—strictly from an analytical point of view, of course. Psychologists studying perception have been fascinated for decades by what they call the "cocktail party effect." Given that name by Colin Cherry in the 1950s, it poses the question: how is it that you can stand surrounded by people in the middle of a party and concentrate on the conversation you're having while ignoring the equally loud talk around you (that is, until your name comes up in one of the other conversations)? If you were to look at a printout of the streams of sound frequencies picked up by your ears, there would be few if any identifiable clues as to which represented any one voice as opposed to all the others. Yet we are highly accomplished at picking out and attending those conversations that mean the most to us.

Some of the cues are obvious: different voices arrive at our ears from different directions and different distances, those voices can be male or female, have different pitches, speeds and timbres, and in a pinch we can glance at the person whom we suspect as the source of the most interesting conversation and read his lips, at least crudely. Equally important is the fact that each stream of speech has an identifiable cadence, flow and continuity of content that allows us to link segments together.

However, even when all those clues have been accounted for, it is still mysterious exactly how we are able to cut through the chatter and follow a single line of speech among many. Cherry and his colleagues repeatedly made the point that having two ears was absolutely crucial. Because sound travels fairly slowly, it's easy to judge the direction from which a sound is coming by comparing the time that sound takes to reach each ear. If it's coming from the right side, the right ear gets it first. At the same time, even though the signals reaching both ears are being compared, our subjective impression is that there is only one sound. As Colin Cherry himself said, "One of the most striking facts about our ears is that we have two of them—and yet we hear one acoustic world; one voice per speaker."

Of course direction is only one issue. A cocktail party inevitably poses the problem that more than one interesting voice is coming from the same direction, and it would be nice to disentangle the speech signals of each. Attention is hugely important: we have to be able to direct our attention to one or another sound easily. One experiment that demonstrated this perfectly had participants listening to two completely different voices, one in each ear, but paying attention to only one. The attentional disparity is so great that people are only dimly aware that there is some sort of noise going on in the ear they are ignoring, but they're not even sure it's speech. Oddly enough, they have little memory of the meaning of the speech they *are* listening to.

It has been suggested that this is what underlies what's been called the "what-did-you-say" phenomenon. Someone says something to you when you're really not paying attention. Almost reflexively you ask, "What did you say?" but after a moment's thought, you realize you did know *exactly* what that person said. Somehow you're able to dredge it up from its somewhat precarious existence on the fringes of your consciousness.

Scientists have long been interested in trying to create a machine version of the human brain's ability to cope with the cocktail party, but they are still far from building such a piece of technology. But while the best-known version of the cocktail party problem is this unravelling of the acoustic signals, another analysis, this one by a physicist, explains why and how such parties become loud.

Physicist William MacLean published his "On the Acoustics of Cocktail Parties" in 1959, but, like all good science, it's just as relevant today. MacLean begins by pointing out that the sound you hear at a cocktail party has two sources: the direct sound coming from the person who's speaking to you and the indirect sound that arrives at your ears only after bouncing off the walls, the furniture, the people and even the cocktail sausages and salmon sandwiches.

Consider a party where there are several conversational groups, and in each a single person is talking. (MacLean limits his analysis to parties with only "well-mannered" guests in attendance—that is, those who won't try to outshout someone who's already talking. On second thought, this might be where his analysis *is* dated.) As a listener, you're hearing the direct speech of the person in your group and also the indirect speech from everyone else in the room who's talking. The party will remain at a manageable noise level as long as, in each group, the speaker can be heard above the background indirect noise, or putting it in physics terms, the "signal-to-noise ratio" remains above that required for intelligible conversation.

But as soon as the indirect sound exceeds the direct sound, there will be trouble. Equations show that the number of guests is the critical factor: as that number rises, the number of speakers rises, and the background noise goes up, until each speaker can no longer be comfortably heard. At that point, the speakers might step up their volume a little, but MacLean's equations clearly show that this would be completely fruitless, because every speaker will do the same, and the background noise will just keep pace.

I should caution you at this point that there is no one magic number of guests for all parties—every party room is a different size and shape, and the reflection of sound from curtained walls, or the absorption by three-piece worsted suits, will be different in every situation. Nonetheless each party *will* have its own magic number of guests, and once that number is exceeded, individual conversations will gradually be drowned out by the background noise. Even if there's a sudden cessation of talk—as when the host calls for silence to toast the guest of honour—it won't last. As soon as talk begins again, it will get out of control exponentially, that is, not just steadily, but at an ever-accelerating rate. Wedding receptions are perfect experimental settings to observe just how fast the sound level returns to its original loudness.

There is only one correction that can be made if the noise level is unmanageable. Speakers and listeners will move closer to each other, the new distance being dictated exactly by the signal-to-noise ratio. The key here is that the sound level reaching your ears from the person speaking directly to you varies in proportion to the square of the distance between you. (For instance, moving three times closer to the speaker will increase the sound level ninefold.) Thus a relatively slight reduction of that distance boosts the sound level significantly, and it's that distance that's critical, not how loud you talk. In fact, although long-term observation would be necessary to confirm this, conversational groups probably contract and expand as the numbers of people at a party rise and fall. If you knew the exact acoustics of a particular party room, you might even be able to calculate the number of people at the party simply by measuring the distance between people!

This is an admirable study, and a perfect example of how concepts from physics can be applied to real-life situations. Of course, as more and more people join the party, partygoers will be forced to crush ever closer together to be able to hear their speaker. In the physicist's model of a cocktail party they might do that, but in the real world they have other worries holding them back: crowding too close together might mean they would miss those juicy tidbits overheard from nearby conversations. A reminder that even physics has its limits.

Déjà Vu All
Over Again

The title is probably Yankee catcher Yogi Berra's best-known quote. And that's saying something, considering he also is credited with saying, "You can observe a lot just by watching," "It ain't over till it's over," "You better cut the pizza in four pieces because I'm not hungry enough to eat six," "When you come to a fork in the road, take it" and "The future ain't what it used to be."

You'd swear Berra had a writer. (In fact the famous baseball commentator Joe Garagiola said more than once that Berra was not the source of all these quotes attributed to him. Whether the author was Garagiola himself or someone else isn't clear.) One thing is sure: none of the scientists who actually research déjà vu turn a phrase as well as Yogi (or whomever). The standard definition of déjà vu among those experts is "any subjectively inappropriate impression of familiarity of a present experience with an undefined past." We'd probably say, "I feel like I've been here before."

If you've had a déjà vu moment you know that they are incredibly vivid. It's not a vague feeling that something is familiar; it's a sudden rush of recognition that you have been in this place before, perhaps even with the people that accompany you now, sometimes even saying the things that you have just said. It's also true that it is one of the things that people ask me about most, especially in conversations about the brain. Once you've experienced it, you're dying to know what is going on to produce a déjà vu.

I'll warn you now: there is no hard and fast answer to that question. That is perhaps regrettable, but on the other hand there are so many theories that you can pick the one you like the best. But before we get to those, here's some of the natural history of the phenomenon.

Déjà vu has likely been around from the time our ancestors acquired modern minds. The first unambiguous references to it, at least in the scientific literature, appear in the nineteenth century, but I have no doubt that people were experiencing it earlier than that. One thing is for sure: once its existence was acknowledged by psychologists and psychiatrists, it became clear that interest in déjà vu was deep and widespread.

There have been countless surveys of the incidence of déjà vu. One thing they make clear is that it is very common: if you average all the reliable surveys (and even throw in some that are dubious), it looks like about two out of every three people have experienced it at least once. There is a curious age distribution though: there aren't many teenagers who report it (but those who do claim to have numerous experiences); the incidence then rises rapidly through the twenties and thirties, and then slowly starts to decline. Even if you had several déjà vus when you were young, you're likely to have fewer as time goes by.

And how many are you likely to have? That's a slippery question, with the answer depending on how exactly the question is asked. Ask people how many they've had and you get one answer. Ask them whether they think such occasions are rare, occasional or frequent and you get another. If you put all the studies together, it seems that about 10 percent of the population claim to have déjà vu episodes weekly! But the numbers are so disparate that this is at best a rough guess.

There are no significant gender differences, but déjà vu does seem to be more common among better-educated people. It also seems to be more likely to happen to people who travel extensively, possibly because they are continually finding themselves in new settings that might trigger the feeling, or because fatigue and stress seem to bring on déjà vu as well, and both could be associated with travelling.

Where are they most likely to happen? Indoors rather than outdoors, more often than not in a familiar setting, although the most intriguing fact of all about déjà vu is that an unusual number seem to have occurred in or around castles. Dunnottar Castle in Scotland, Pevensey Castle in Sussex, Albrechtsburg at Meissen, Germany, and several others have all been the scenes of dramatic déjà vu experiences. No one knows why, although it's obvious that they are spectacular and somewhat spooky places—whether that tips you toward déjà vu isn't really known.

It's odd that castles would be special places, but the idea that a place would enhance the chances of déjà vu isn't surprising. Surveys put locations as the most important environmental feature, followed by the words of others, your own words, the actions of others, your own actions and, finally and least important, objects.

Getting any kind of solid data about déjà vu is difficult because the experience is totally subjective. I might see that you're startled by your déjà vu experience; I can be impressed by the intensity with which you appear to be reliving that moment, but in the end I have no idea exactly what you've been through, or even if your claim to have had a déjà vu experience is reliable. That's one problem. There are others, one of which is that questions about déjà vu have often been imbedded in surveys dealing with other mental experiences, especially those of the paranormal kind. Clearly, if you ask people whether they believe in ghosts, telekinesis, ESP, witches *and déjà vu,* their answer is likely to be congruent with their views on the paranormal, whether positive or negative.

While we're on the subject of the paranormal, it's worth noting that besides the scientific approach to explaining déjà vu, which I'll get to shortly, there are other ways of viewing the phenomenon, one of which is indeed through paranormal-coloured glasses. The obvious connection to strange otherworldly possibilities would be that a déjà vu experience reveals another life you have led in the past. The obvious problem with such a thought is that where on earth—or elsewhere—would you have led a life where you were standing in front of a home theatre talking to people about Harry Potter?

Perhaps a bent toward the paranormal coupled with the realization that previous-life explanations just didn't cut it has led to variations on this theme. One is precognition, the idea that you know when something is about to happen. Apparently having a déjà vu has convinced some people not only that they have been there before but also that they know what's happening next. One of the scientific explanations for déjà vu (an idea that there are two paths for incoming information and one might be unusually delayed vis-à-vis the other) can, admittedly with some difficulty, explain why precognition might seem to be happening.

Of course it is also possible that when you experience déjà vu you are momentarily living someone else's life, someone who was in that place either in the past or the future. Of course it is also possible that the life you're experiencing is your own and that you were indeed in this place, or a place like it, and only remember a few fragments of that experience, enough to trigger familiarity but not enough for recall. That too would be an offshoot of the scientific explanations that are forthcoming.

Alan S. Brown of Southern Methodist University in the United States has pretty well cornered the market on the science of déjà vu, having written both a recent scientific paper and a book on the subject. He points out that besides the paranormal, there has been a wealth of what he calls "psychodynamic" explanations of déjà vu—that is, explanations with a heavy dose of psychiatric wisdom. Brown has little patience for either, claiming not only that they fail to lead to useful approaches for future research but also that they turn off scientists unwilling to work in areas heavily influenced by thoughts of reincarnation or wish-fulfillment.

Nevertheless it is true that psychiatrists have had a field day with déjà vu, although they seem not to have come to any significant agreement. Sometimes the disagreement occurred within the individual. None other than Sigmund Freud changed his mind several times about déjà vu, claiming first that it represented unconscious fantasies, then that it was an anxiety reaction to a castration fantasy, then that it was depersonalization and finally concluding that it was just too complicated to categorize.

Brown quotes the following passage in his book as an example of why you might want to steer clear of psychodynamics:

> The déjà vu thus involves a controlled regression in the service of the ego as a consequence of the defensive and frantic search of the ego for the symbiotic and nonsymbiotic good, omnipotent mother, rapidly scanning the phases of life in a descent historically to the composite primal-preobject-early libidinal object-representations of mother.*

After something like that the science is easy, but there *is* plenty of it. A list of the myriad theories of déjà vu would tire you—and me—quickly. Instead I've selected a few of my favourites. They are hunches more than theories, there's not a huge amount of data to support any of them, no one of them is the "best" explanation, but they do give you the flavour of the science of déjà vu.

In contrast to both the paranormal and psychodynamic theories, the scientific approaches are all based on mechanisms that either are known to exist in the brain or at least should be identifiable even if they haven't been identified yet. These ideas have a material basis. One approach is to say that two processes in the

---

* Pacella, B. L., "Early Ego Development and the Déjà Vu," *Journal of the American Psychoanalytic Association* 23 (1975): 300–318.

brain that normally operate together suddenly have that close relationship disturbed, with déjà vu as the result. An example is the dual process of recall and familiarity. Usually when you remember something, familiarity is part of the package. But if that tight relationship is momentarily lost, it would be possible to feel that a situation is familiar but be unable to remember why.

There could be other twinned processes that momentarily go awry. Psychiatrist Andre de Nayer proposed a tape-recorder analogy: he likened memory encoding and retrieval to the record and playback heads of a tape recorder. Usually only one is functioning at any time, but if by some fluke both memory encoding and retrieval happen at the same time, that makes the newly formed memory somehow seem familiar: déjà vu.

It could also be that the two processes that come unhinged are pathways for incoming sensory information, one faster than the other. Normally there is no more than a split-second difference in arrival times for the two, but imagine if, for some reason, that time difference is unusually stretched out. A strange perception results. The fast message arrives and is processed; then the slower one arrives and is perceived not as the delayed version of the same message but as a completely different one. One that is strangely similar to the one that is already in the memory banks.

These are scientific theories, but you no doubt noticed that they aren't exactly specific, and in fact, they can't all be right. Those three are so different that, at best, one might be an accurate representation of what's going on.

Distortions or failures of memory make up a whole different group of explanations for déjà vu. One such theory argues that all you really need for déjà vus is to have one small part of a scene be something familiar, and that familiarity will bleed over to colour the entire scene. Alan Brown cites the example of entering a strange room and seeing a grandfather clock just like the one in your aunt's home. If you can't identify the cause of the familiarity feeling, you're likely to attribute it to the entire situation. It doesn't have to be just one element either. One researcher, after thinking hard about his déjà vu experience, was able to identify three things in it that were memories that he knew he held.

Finally, what might be the most persuasive explanations for déjà vu involve attention and distraction. Imagine you're walking down a country road, attending casually and absent-mindedly to your surroundings, when your full attention is suddenly captured by a bird flying right by your nose. That shift of attention has for all practical purposes wiped away whatever memory you had of the surroundings, which, when you return to your idle survey, seem oddly familiar as they come into attentional focus, as if you were just there. (You were!)

Attention is much more important than we realize. Try listening carefully to a CD in your car when traffic is busy. It's harder than you think. Brown suggests the experience of walking into a room while talking on your cellphone and looking at the room but not noticing, not attending to, some "stimulus," say a pot of flowers. Then you hang up the phone, switch your attention to the room—and the pot—and have a fabulous déjà vu experience.

So take your pick. I like the attention and memory versions, but they are all viable. No definitive answers yet, but lots of good ideas. And I can't leave this chapter without talking about two other related experiences. One is the reverse of déjà vu, "jamais vu," the feeling that you have never been in a place that you know should be completely familiar. There aren't any really good explanations of jamais vu, although one tentative one is that it could be something, even small, that might be different in the familiar setting. So even though the room is largely unchanged, one chair has been moved, and that single change could trigger the overall jamais vu feeling.

But more intriguing to me is a related phenomenon that all of us experienced when we were kids. Repeat a word over and over and eventually it becomes meaningless. If you reach this state by hearing a spoken word over and over, it's called "semantic satiation." If you accomplish the same thing by staring at a single word for a minute or so, that's "word alienation." Both are altered forms of jamais vu. And again, as is the case with both jamais and déjà vu, we are still largely in the dark as to why they happen.

Sweet Thoughts from a Tiny Brain

No common food is taken more for granted than honey. Do you realize that it takes the combined brainpower of thousands of females to get that honey to your table? It's not that the packaging and merchandising of honey is a labour-intensive operation employing mostly women—those thousands are the worker honeybees who gathered the nectar that was processed in the beehive to make the honey. The energy they expend in flying from hive to flowers and back, dozens of times a day, is impressive enough, but the mental work involved in nectar gathering is what is really startling.

A worker bee has a brain the size of the ball in a ballpoint pen, about one one-thousandth of a gram of grey matter. Our brains weigh a million and a half times as much. But even with this seemingly crippling deficit in brainpower, bees perform what seem to be very complicated feats of memory and communication.

One of their secrets is that they don't second-guess themselves very much. Bees' brains are programmed to deal only with very specific kinds of information, and then to act automatically and inflexibly. They are, in the language of the computer age, minia-turized flying robots equipped with a nectar-gathering software package. Keeping things simple and repetitive is a common strat-egy among insects, and it was first dramatically demonstrated by the great nineteenth-century entomologist Jean Henri Fabre. He was studying a wasp that would carry a paralyzed cricket to the door of its den, then reconnoitre inside before coming back out to haul the cricket inside to lay her eggs on it. One day Fabre moved the cricket back very slightly while the wasp was inside. She returned, found the cricket out of place, moved it back to where it belonged, then returned inside to check out her den. Fabre moved the cricket again, and the same scenario ensued. After forty repe-titions Fabre tired of the game, but the poor wasp, irretrievably stuck in her own simple routine, would have gone on forever. She couldn't think her way out of the predicament—in fact, she wasn't even aware she was in a predicament!

Foraging bees, too, can be revealed as unthinking automatons like Fabre's wasp. By the time a honeybee goes on her first flight searching for flowers, she's probably already three weeks old, but experience is not necessary.* She is programmed to check the colour, shape and odour of the flowers she lands on, so that she will remember any that are particularly rich sources of food. But

---

* She has been working in the hive doing all the other things worker bees do: tending the brood and the queen, ventilating the hive (by buzzing her wings) and performing guard duty. Foraging is her last act—four or five days of it, and then she'll die. Oddly enough, not of old age but of distance. Some studies have suggested that foraging bees have a few hundred kilometres of flying in them, after which they die, whether that takes five or fifteen days.

she doesn't waste her time scanning the entire countryside as she flies over it. Her brain is wired to record flower information in very narrowly defined and efficient ways. For instance, she records the colour of a flower only in the last three seconds of flight before she actually alights on it. Any colour she's seen prior to that is unregistered, and any new colours she sees while standing on the flower make no impression on her. And she has to be flying to record the colour—if you hand-carry a bee to a flower, she'll never remember what colour it was.

She also learns what time of day each variety of flower produces nectar, so she doesn't waste foraging time and energy. Scientists have been able to train bees to discriminate between nine different times of day, some of them only twenty minutes apart, but again, a little experimenting reveals how rigidly the bees obey their own time clocks. If a blue tray of sugar solution is put out from nine to ten o'clock in the morning, then is replaced by a yellow one from ten to eleven, the bees will feed from the first tray, then the second, and in the process learn to associate time with colour. The problem is that if the first tray is not removed, the bees will persist in switching to the other coloured tray at ten o'clock, even if the first tray is still full of sugar. They're locked into the simple routine, "blue at nine, yellow at ten."

Finding and remembering flowers is only the beginning: it's when foragers get back to the hive and tell the others what they've found that the bees' brains are really put to the test. The incoming forager tells the others where the good flowers are by dancing. The direction, orientation and intensity of her movements tell them the distance, direction and quality of the flowers that she has visited.

The most famous dance is called the "waggle dance," performed when the flowers are at least a hundred metres from the hive. The dancer stands on the vertical face of the honeycomb, in complete darkness, and starts to walk in a figure-eight pattern on the surface of the comb, waggling her abdomen furiously as she goes. The other bees cluster around her, touching her as much as possible. From the speed and size of the figure-eights she dances, the number of waggles and the orientation of the dance on the honeycomb wall, they can infer the distance and direction to the flowers.

This is pretty amazing in itself, translating her memory of the trip to the flowers into dance movements in a dark hive, but there's much more to it than that. For one thing she's not actually dictating a specific distance to her hive-mates but rather an estimate of the amount of energy they'll expend to get there and back. So she'll perform different dances for the same flowers on calm and windy days. This has been proven by loading bees with tiny weights or flaps that produce air drag—a bee so encumbered dances to indicate a greater effort than she did when flying free. Bees that were forced to walk (!) to a nectar source indicated a greater distance yet. They even differentiate between a headwind going out and a headwind coming back, because on the way back the bees are carrying the extra load of nectar. Recruits take this message seriously: the harder they're going to have to fly—as indicated by the dance—the more honey they take in as fuel for the flight.

Even the fantastic detail conveyed by the dance could be explained by assuming the bee uses automatic brain programs engineered by millions of years of evolution. But in the last few years, some experiments with bees have revealed abilities that need new explanations. For instance, biologist James Gould at Princeton tricked bees into dancing to indicate a source of nectar right in the middle of a lake. The dance audience in the hive paid close attention, flew quickly to the edge of the lake, then patrolled along the shore, apparently searching for the supposed flowers. They refused to go to the spot in the middle of the lake indicated by the dance, even when there was a boat loaded with nectar floating there. Gould proved they weren't just afraid of the water by having scouts dance to indicate a source on the bank on the other side of the lake. Recruits didn't hesitate to fly across the lake and collect the nectar. If this were just automatic behaviour, why didn't the bees try looking in the middle of the lake?

This is not an easy one to explain. Some biologists think that bees may retain in their memories a series of snapshots of the terrain over which they've flown—a "neural album"—which they can shuffle through mentally and plot where they are. If that's true, their refusal to fly to the middle of the lake suggests the bees could first match the locale indicated by the dance to a snapshot of the lake (presuming they had one in their collection), then refuse to forage there, knowing that lakes don't have flowers.

James Gould suspects that rather than an album of snapshots, bees have a mental map that they can refer to just as we do when driving. Bees with maps in their brains would compare the dance information to their map and see that to follow it would place them in the middle of the water, apparently an unacceptable conclusion. In another experiment, Gould trained bees to fly from their hive to a feeding station that was surrounded on all sides by trees. He then transported them to a completely different site, from which they could see the hive but not the feeding station. He found that most of the bees did not go back to the hive first, then to the feeder, but flew directly to the feeder, a flight that they had never taken before.

How do bees manage this? One clue is the orientation flight, a flight that workers about to become full-time foragers undertake before they actually start searching for sources of nectar. The flight starts with the bee hovering in front of the hive entrance, facing it, and flying in short arcs back and forth. Gradually she'll extend those arcs into larger circles, and finally will spiral her way up and out of sight. But she will return in a few minutes without nectar. And it's not just the lack of nectar that reveals this is a different kind of flight: foraging bees never pause to look back at the hive, but take off straightaway. Unfortunately no one really knows what goes on in the bees' brain during these orientation flights, but it's clear they are laying some sort of foundation for the navigational feats of which they are eventually capable.

The idea that bees might have mental maps is a little hard to swallow, but even maps can't explain the mathematical abilities of bees. If you put out a sugar solution a few metres from a hive, pretty soon you'll get all kinds of bees feeding there. Then the next day put out the same dish 25 percent farther away. Again the bees have no trouble finding it. Then repeat the same routine, each day moving the dish farther by 25 percent of its previous distance. While you might have been increasing the distance from the hive a metre or so at first, by the end of a week you'd be adding several metres each time. Not only do the bees have no trouble finding the nectar in this experiment, they will actually start showing up early—in the exact place where you are going to put it! Each day they calculate what a 25 percent increase in distance from the previous day is, and fly to that place. That's known in math as a geometric progression. Bees calculate it using a brain that is a cubic millimetre in size. So far no one has come up with a completely convincing explanation of how they can do it.

At one end of the explanation spectrum is the idea that, in some mysterious way, bees are sensitive to quantum effects, the mechanics that determine the properties of the parts of atoms. This claim came from some mathematical descriptions of the bees' waggle dance, but, at least at this point, a description this far removed from the rest of our knowledge about bees seems destined to languish, at least until the biologists catch up. However, those biologists, or neuroscientists, are making strides in understanding how the bee's brain is able to perform the feats that it does. This research shows that the organization of the bee's brain is more sophisticated than we ever thought.

Biologists like Randolf Menzel in Germany and Martin Giurfa in France have been analyzing the bee's behaviour and have shown that its brain must be more than just an assemblage of separate mental modules each of which has one task and one task only. There *are* modules, but they seem capable of communicating with other such modules to put information together in new ways and allow the bees to behave in extraordinary ways.

For instance, honeybees can be trained to choose the right target—say, a circle with a spoke pattern rather than concentric rings—to be able to get a sugar-solution reward. That's not so impressive in itself. But that same bee can be trained to respond to the spoked circle at the entrance to a maze, but also to the ringed version to gain entrance to the nest. In that case, it's all about context.

But it gets even better than that. In certain experiments, if the bees need to access a maze by choosing the spoked circle at the entrance, they are then taught that they must choose it again to get the sugar solution inside. But here's the amazing part. Once they've mastered the spoked/ringed circle test, they can transfer that concept—"I have to choose the circle with the same pattern to get the sugar"—to a pair of gratings with lines running at different angles. It's no longer a case of a rigidly determined choice of a particular pattern; it's the choice of same versus different. And remember, that choice is being made in an exceedingly tiny brain.

The fact that one worker bee may collect only enough nectar for a mere thimbleful of honey—in her entire life of foraging—has led bee fanciers to emphasize the tremendous physical labour that goes into putting that honey on your table. The thought that goes into it is no less impressive.

THE BIG BANG
ON CABLE

The next time you find yourself in front of the television set looking for entertainment, throw away the TV guide and find a channel where there's absolutely nothing on (yes, yes, I know that's most of them), a channel where there's nothing but snow. Or if you're idly scanning through the stations on your car radio, turn to the high end of the FM dial, somewhere around 108 MHz, and listen for the hiss in the space between stations. Once you've found it, you can revel in the unbelievable fact that your radio or TV is giving you a glimpse of the beginnings of the universe.

The snow on your TV set is just electric noise, a disorganized collection of unrelated photons—particles of electromagnetic radiation—flooding the antenna from all directions. Some even come from the antenna itself. Three or four percent of those photons have been flying around outer space for 13 billion years or so, having first appeared a mere 300,000 years after the Big Bang, the explosion that began our universe.

The idea of the Big Bang is not an easy one to come to grips with. You have to start with the idea that the entire universe that we can see from the earth today is expanding: all the galaxies, and indeed clusters of galaxies, the billions or even trillions of stars in each of them, and the gigantic swirling clouds of dust and gas are all rushing away from each other at fantastic speeds. And they're not fleeing to empty corners of the universe—space itself is expanding. The analogy often given is that the galaxies are like raisins in a rising loaf of bread. They're moving away from each other because the dough between them is expanding. Obviously, if when we look out into space today we see everything rushing away from everything else, it's reasonable to suppose that a long time ago everything was closer together than it is now, and the further back in time, the closer together it was. Physics dictates that the smaller and denser the universe was, billions of years ago, the hotter it was. So if you extrapolate back far enough in time, like reversing a film, you come up with the entire universe crammed into a tiny ball that's so hot and dense that it defies modern physics. That ball, however *it* came into being, exploded outward. The explosion is now called the Big Bang, and we're living in its aftermath, the expanding universe.

The universe, having begun as an object that's too hot and too dense for explanation, cooled and expanded rapidly enough that even within the first small fractions of a second, it had assumed a form that physicists can comprehend. Even so, for hundreds of thousands of years it was still too hot for atoms to form, because the moment the parts of atoms, principally the protons and electrons, formed tentative bonds, they were torn apart—they couldn't withstand the agitation caused by the heat.

As long as atoms couldn't form, the particles of light—the photons—were trapped. They'd travel a tiny distance only to be deflected or absorbed by free-floating electrons. Light went nowhere at that time, even though there was lots of it. Astronomers describe this as the era when the universe was "opaque": everywhere bright, but nothing to see.

However, about 300,000 years after the Big Bang, there was a history-making event—the scientific equivalent of "Let there be light!"* The temperature of the steadily cooling universe suddenly reached a point where protons, neutrons and electrons could stick together and form atoms. Once they did, the photons were suddenly free. There were no more loose electrons to intercept them, so light could travel in unimpeded straight lines, as it does today. That's when your TV-set photons got their start.

The temperature of the entire universe at this point was about 3,000 degrees Celsius, roughly the temperature of the filament in a light bulb, and the particles of light that were set free reflected that: they had the very short wavelengths typical of photons released from a hot object. (The most common example of this process is an element on an electric range. As it heats up, its colour changes because the radiation coming from it has shorter and shorter wavelengths. Even when it's cold, it's radiating, but the waves are so long, you can't see them.)

Since that time, 13 billion years ago by current reckoning, a lot has changed. Because the universe has been expanding steadily, the waves of these light particles have been stretched out proportionally. Very short when the universe was relatively small, they're now

---

* These days that date has been refined to about 380,000 years after the Big Bang, an astounding feat of precision science.

very long—wavelengths more typical of the radiation that would be produced by something very cold rather than something hot. That's exactly what astronomers discovered in the mid-1960s: no matter where you look, the universe appears to be filled with the kind of radiation that would be produced by an object that's only three degrees above the lowest temperature possible, absolute zero. That's what has happened to the photons released in that momentous event 300,000 years after the Big Bang: what was an explosion of light that filled the universe has become nothing more than a cold, faint, invisible glimmer.

One of the striking things about this leftover radiation is that it is extremely consistent—smooth—no matter which direction you look. It seemed at first unbelievable that in the vastness of space there would be such homogeneity, but the puzzle was solved by physicist Alan Guth, who realized that if the universe had begun as a tiny—incredibly tiny—point, then exploded at some unheard-of rate, then its original smoothness could have been preserved. That rapid expansion was dubbed "inflation," and it is now one of the most important chapters in the history of the universe.

The cosmic radiation may be only a shadow of what it once was, but the photons continue to rain down on us. It's estimated that 500,000 of them are still left in every litre of the universe, and given that they travel at the speed of light—they *are* a form of light, after all—several hundred trillion of these will pass through the palm of your extended hand every second! There's no risk to you, of course—this radiation has an intensity of only one ten-millionth of a hundred-watt light bulb. But it is still enough to produce a few pops and crackles in our radios and television sets. However you pick up your television signal, whether by satellite dish, cable or just plain aerial, it has spent part of its time travelling through the air in the form of photons. When you tune in a channel, you are selecting photons that are travelling in concert in waves of a specific length. No matter what television station you are tuning in (and on some FM radio stations), some of the photons from the Big Bang have just the right wavelength that a few of them contribute to the radio hiss or the television snow. Right in your living room there is real evidence of the Big Bang.

While this cosmic background makes up only a tiny percentage of the snow or hiss, and though it is admittedly a pale shadow of its original awe-inspiring splendour, it *is* commercial-free and perfectly suitable for family viewing.

I Think,
Therefore I Blink

We blink an average of fifteen thousand times a day, each one lasting roughly three-tenths of a second. That's about an hour and fifteen minutes each day we spend with our eyes partly or completely closed. But it's only in the last few years that scientists have really begun to understand what's going on when we blink. There are some situations where you expect to blink: when dust or smoke gets in your eyes, or when you're startled by a sudden noise or the sudden appearance of something close to your eye. But those account for only a small proportion of the total number of blinks during the day. The others have nothing to do with cleaning or protecting your eyes—they are actually signals of what's going on in your brain.

A blink in detail is a remarkable event. As light as the eyelid is—it's the thinnest piece of skin in your body—it doesn't get up to maximum speed immediately. Slow-motion replays show that the eyelid begins to drop, builds up speed to a maximum, then begins to slow again before your eye is actually closed. All of that takes about one-tenth of a second. The eyelid stays closed for about one-twentieth of a second, then it starts accelerating back upward again, leaving a film of tears behind. The odd thing is that even though your eye is partly or completely closed for three-tenths of a second or more, you aren't aware of missing anything. Yet if the room lights are shut off for a much shorter length of time, you notice it immediately.

An ingenious experiment revealed why we don't experience a little blackout every time we blink. Researchers placed a fibre optic cable against the roof of the mouth and shone light on the *back* of the eye, so that the closing eyelid wouldn't interrupt the light signal. They found that the sensitivity of the eye to a flash of light starts to drop just before a blink begins, and stays low until after it has finished.

More recent elaborations of that experiment have shown that it is really not just the eye's sensitivity to the blink, but the brain's. Using the same technique of shining a laser light on the back of the eye to ensure that the retina would be bathed in light whether the eye was open or closed, scientists at University College London showed that even parts of the brain far removed from the primary visual areas responded to a blink. Some of these are areas known to be involved in awareness, areas that have to be active if you are to notice something.

What seems to be happening here is that when the brain gives the order to the eyelid muscles to contract and begin the blink, it also alerts several other areas in the brain to suspend operations for a moment. As they do, your visual awareness is shut down, and you have no idea that your eyes have just closed. This is why you're never aware of the interruption caused by a blink, but when the room lights flicker you notice the change immediately because eye and brain are still operating at peak sensitivity.

Of course, not only do you not notice the darkening but you don't notice any interruption either. The scene you're watching continues to unfold smoothly. That's possibly because this suspension of operations includes brain cells whose job is to alert the brain to changes in the visual world. In the absence of that alert, what was seen just before the blink might linger in the mind to fill in the blank.

(There were hints in this latest experiment that the same sort of widespread shutdown may also be involved in another visual trick. When our eyes are wide open, they do not stand still. They are constantly roving the scene before us, flicking from one interesting feature to the next. Not only are we ignorant of these movements, we are completely unaware of any blurring or shifting that you'd expect would be caused by them.)

Most of us blink about fifteen times a minute, yet apparently only one or two of those are necessary to keep the surface of the eye lubricated. In some forms of Parkinson's disease, patients blink only once or twice a minute, yet have no problem with dryness of the eyes. In fact, scientists Eric Ponder and W. P. Kennedy of Edinburgh University showed in the 1920s that it made no difference to the frequency of blinking whether subjects were in the extremely humid conditions of the botany department's hothouse or in the extreme dryness of a Turkish bath.

Ponder and Kennedy also tried to determine if there was a sex difference in blinking. They found that on streetcars men blinked every two and a half seconds, while women blinked every six seconds. But curiously, the results reversed in a library: while the women again blinked roughly every six seconds, the men had reduced their rate dramatically to one blink every eleven seconds. You somehow get the impression from Ponder and Kennedy's explanation of these observations that Scotland in the 1920s was not yet tuned in to the feminist movement. They describe the women on the streetcars as exhibiting an "obvious lack of interest in their surroundings," while the men are peering out the windows and at each other. Paying attention to the ever-changing outside world always correlates, the authors say, to a higher blinking rate. In the library the roles are reversed: males are paying more attention to their reading, while the women's attention is wandering. In fact, Ponder and Kennedy suggest peevishly that the females moved their heads and eyes so much it was difficult to count their blinks.

Regardless of their interpretations, Ponder and Kennedy's observations set the stage for the modern investigation of blinking. As it turns out, the sex differences they measured have never shown up again in controlled situations, but the two scientists hit on something important when they showed that different kinds of mental activity are accompanied by different rates of blinking. Research since then has boiled that relationship down to one simple rule: the harder you are concentrating, the less you blink.

This has been demonstrated in all kinds of circumstances: pilots in flight simulators cut their blinking in half when they move from the co-pilot's seat to take control of the aircraft. If the simulated flight is low level and high speed, both co-pilot and pilot blink even less. Car drivers blink less in city traffic than on highways, and don't blink at all as they're passing trucks at high speed.

Even in situations where there's no apparent hazard, concentration reduces the number of blinks. Doing mental arithmetic or remembering series of numbers reduces blinking. If you are conversing you blink at the normal rate, about fifteen per minute, but that drops to six a minute if you're reading. (So much for the idea of intense conversation.)

A closer look at reading begins to reveal what blinking really means. Studies more than forty years ago showed that readers blink most often when they reach a punctuation mark or the end of the page. In other words, when there's a pause in the flow of information coming into the brain, there's a blink. A blink is a visible signal that the brain is taking a breath.

It's not just with visual activities like reading. Subjects who are being quizzed by researchers blink very little during a question and in the short period immediately following, presumably because they're thinking about the answer. As soon as they start to answer, they blink, sometimes repeatedly. In another experiment people were required to distinguish between a long musical tone and a short one. They usually blinked sometime after the end of the short tone, but often right in the middle of the long tone. The explanation is simple: when the short tone ends, there is a brief delay before the brain realizes it's over. It then recognizes that fact with a blink. But as soon as the long tone stretches beyond a certain length, the brain knows it must be a long tone, and blinks even before it's over.

So a blink appears only after the brain has processed a certain amount of information, and the greater the amount of information, the longer the delay before the blink. If it takes twice as long to memorize six numbers as two, then the blink will appear that much later. Blinking patterns may even reveal different thinking styles. Some people doing mental arithmetic blink only once, at the end of the calculation—they are the equivalent of the driver who doesn't blink as he's watching the road. Others blink repeatedly as they calculate, perhaps because they're solving the problem one step at a time. (A parrot was observed to decrease its blinking rate from once every 17 seconds to once every 25 seconds while listening to a foxtrot—the significance of this is not yet known.)

Why should blinking reflect what's going on in our brains? Apparently it's not necessary: those same patients who blink only a couple of times a minute think perfectly well. Blinking isn't a cause of the thinking process, it's a result. Some psychologists think that blinking just represents a spillover of brain activity. The flurry of nerve impulses produced during reading or thinking somehow escapes into the nerves controlling the eyelid muscles, and those muscles twitch. It's also true that people who are asked to recite the alphabet to themselves or to silently count upward from one hundred blink much less often than if they're asked to do the same things aloud. Somehow the act of speaking (without changing the thoughts involved) increases the rate of blinking. This supports the idea that there's some kind of spillover, and in fact the area in the brain that directs the movement of the eyelid muscles is immediately adjacent to the area controlling the tongue and face.

Most people believe that anxiety increases the rate of blinking—former U.S. president Richard Nixon blinked twice as much as normal when answering hostile questions about Watergate as he did when answering non-threatening questions, and during the

1988 television debates between Michael Dukakis and George Bush, both debaters blinked more when questions were directed their way. But these findings are confounded by the fact that thinking and talking both increase the rate of blinking anyway. Ponder and Kennedy, the same researchers who in the 1920s witnessed blinkers in streetcars and libraries, did note that witnesses in court blink more rapidly—sometimes twice as fast as normal—when they are being cross-examined. But more recent studies showed that subjects who were measurably anxious (because they had been threatened with electric shocks) did not increase their blinking rate. And even if anxiety plays some role in increasing blinking, no one has any idea why this should be so.

The other common belief about blinking is that it reveals a certain timidity in the face of a threat. Two enemies stand toe to toe, the first to blink is the loser, and the winner proclaims, "He blinked." The phrase is used to describe confrontations between individuals or superpowers, and I think its origin is clear. An exhaustive survey of blinking in animals revealed that, on average, carnivores blink less often than herbivores. Or, if you want to put it another way, the predator blinks less than the prey. One explanation offered in the 1940s was that the prey needs to survey the landscape continuously for danger, is therefore changing focus and direction of gaze constantly, and blinks with each shift of attention. The carnivore, on the other hand, must fix its gaze on its prey—without a blink. Unfortunately, the victors in the human versions of these situations, while crowing that the opponent blinked, forget that an unblinking face fronts an unthinking brain.

Dog Daze

The last twenty years have seen an explosion in research on our canine companion. Most of it is straightforward—if surprising—behavioural work, like studies that show that dogs are actually better than chimpanzees at picking up cues from humans (so exquisitely tuned to us they are) or experiments that demonstrate that dogs somehow know when other dogs are paying attention to them and alter their behaviour accordingly (revealing they have at least some ability to understand what other animals are thinking).

These are all fascinating, but I admit that I have a soft spot for research that is a little more off-the-wall, a little outside the normal bounds of doggy research. There are two such examples, each a question: Do dogs look like their owners? and, Do dogs know when their owners are on the way home? The first is slightly more respectable than the second.

The folk belief that people look like their dogs is so prevalent that it became the theme for a series of national advertisements for the telephone network Fido a few years ago. But scientists have actually determined that there is some truth to the idea. In 2004 two independent groups of researchers, one in Argentina, one in California, simultaneously ran experiments that supported this claim.

Here's how they worked: the scientists gathered pictures of dogs and their owners and presented them to volunteer participants. The volunteers had to match the dog to the owner. But the set-up can't be as straightforward as that; there has to be some organization so that statistical meaning can be drawn from the results. The California group, Nicholas Christenfeld and Michael Roy, had their subjects view one owner picture and two dog pictures and judge which dog belonged to the owner. Christenfeld and Roy chose the combinations so that every owner/dog pair was seen with each of fourteen other dogs. The pictures were taken in dog parks, but the researchers were careful to have owners and their dogs photographed with different backgrounds, ensuring that judgments would be made solely on the appearances of the human and the dog.

The results, while not spectacular, were good enough: sixteen of twenty-five purebred dogs could be matched with their owners, a result significantly better than chance. There were some subsequent academic protests that the experiments were flawed, but they have stood the test.

The results raise some immediate questions. How do people end up with dogs that look like them? One possibility is that dogs and their owners could grow to look like each other over time, in the same way that married couples do. The most (and practically only) quoted study on this was published in 1987. In it, photos of husband and wife—at their marriage and again after twenty-five years of married life—were compared to random pairings, and volunteers were asked to decide which pairs were married. Again, they were able to do so more accurately than if they had just been guessing. The authors suggested that couples who had lived together for a long time would start to look like each other because they had experienced all the great emotional moments together: the birth of a child, their first house. According to a theory called "emotional efference," emotions, blood flow in the face and brain, and facial musculature are all intimately related. Different emotional experiences are accompanied by different movements of the face. Over time, the authors suggested, the settling of the facial muscles into habitual facial expressions alters the topography of the face, and does it simultaneously in both partners.

This may not be the only possibility. It could also be that years of actually imitating the facial expressions of one's spouse could start to make two faces look like one. But however fascinating this idea is, it seems not to apply to dogs and their owners. In their paper Christenfeld and Roy (who unfortunately sound like some sort of Vegas act) established that there was no link between the resemblance of human and dog and the length of time they had been together.

The only conceivable alternative is that people choose dogs that already look like them (although of course with some breeds, particularly those that have a reputation for toughness, it's possible that some people buy them just for their behavioural resemblance, not their facial similarity). You might have noted that it was purebred dogs who matched their owners; mutts didn't. A reasonable explanation for this would be that an owner looking for a match would go purebred because he or she would know what the dog would look like as an adult. Mutts aren't so predictable.

And why would anybody do this? What kind of foolish human behaviour could this be? The other research group to consider this, Venezuelan researchers Christina Payne and Klaus Jaffe, argue that this is a spillover from a well-documented phenomenon called "assortative mating." Assortative mating describes how males and females of a wide variety of species look for mates who are somewhat similar to them. Not exactly like them, but not completely different either.

Assortative mating is really a genetic strategy that ensures that the offspring of mates who resemble each other will represent the ideal combination of genes and so will carry the family banner well into the future. Given that it is widespread throughout the animal world, it must also have a long genetic history. Payne and Jaffe wonder if the same sort of urge to seek out a mate who looks the same might extend—unconsciously—to the search for a companion. A canine companion.

Well, who knows? There is lots of evidence that mate-seeking animals operate according to the rules of assortative mating, but there is really no direct evidence that such rules can affect other non-mating behaviours. But the evidence of similarity between dog and owner is there and has to be explained.

Remember, none of this is necessarily conscious, and probably isn't. We evaluate many other characteristics of people unconsciously, like their waist-to-hip ratio and degree of body symmetry, so why not facial similarity? Of course it is only one element in a whole mess of influences on even the simplest decisions. But there is no reason to doubt the existence of something like this just because you aren't aware of it.

So the answer to the first question is, Yes, dogs *do* look like their owners. Question two is, do they also know when their owners are on their way home? Is the affiliative bond between dog and human so strong that it can produce behaviour that looks, at first glance, extrasensory? Some say yes. Some say no. Unfortunately what seems to be happening is much more intriguing than the analysis of it. But if claims like these don't hold up to scrutiny, then they have to be abandoned, whether dog lovers like it or not. (I'm a dog lover, but my dogs either don't know that I'm on my way home or don't care.)

The maverick English scientist Rupert Sheldrake, best known for claiming that he has gathered various kinds of evidence for telepathy, argues that his studies show that dogs can tell when their masters are on the way home, even when those masters are still much too far away to be detected by any known animal sense. Dogs see pretty well, hear very well and are extraordinary smellers, but none of these senses would seem able to identify a car that's still several blocks away. But that's what Sheldrake argues.

His most famous case involves a dog named Jaytee and his owner, Pamela Smart. Jaytee was a terrier crossbreed who, Pamela's parents reported, would go out every day and sit at the window shortly before Pamela came home. As long as Pamela was employed, she would leave work at about 4:30 and Jaytee would begin his vigil at about that time. Later, when she no longer had that job and was travelling to and from the house at irregular intervals, Jaytee still seemed to sense when she was on her way home. Sometimes he would go to the window at about the time she set out for home; sometimes he went on the alert early.

Jaytee apparently did this so regularly and consistently, not to mention mysteriously, that he attracted wider attention. In 1994 an Austrian television crew simultaneously videotaped Jaytee in the house and Pamela as she walked around town. The tape was striking: the moment Pamela is informed by her camera crew that the randomly selected time to return home was upon them, the camera monitoring Jaytee revealed that he pricked up his ears, then within seconds went and sat in the window. At exactly the same time, Pamela was walking to a taxi stand to get a cab home.

By this time Rupert Sheldrake was involved in the story, and together he and Pamela collected reams of data on Jaytee's behaviour that showed, to their satisfaction, that this dog was somehow aware of his owner's imminent return home, or even the simple intention to do so. They published their first analysis in *Journal of the Society for Psychical Research*. In that report, besides presenting the supporting data, they considered some possible and unremarkable explanations for the behaviour: that Jaytee somehow heard the familiar noise of Pamela's car, or that he was tuned to the fact that she returned home at about the same time every day, or that he picked up on the anticipation of Pamela's parents (who took care of him while Pamela was out) or even that he just kept going to the window over and over until she arrived.

But none of these seemed to be an adequate explanation. Pamela often took unfamiliar cars, like cabs, home; she was sometimes too far away to be heard anyway; her parents usually had no idea when to expect her, and Jaytee's trips to the window (with the exception of some that had obvious causes, like the sound of another dog) appeared to be aligned with Pamela's setting out for home.

However, skeptics took a different point of view. English psychologist Richard Wiseman, in co-operation with Sheldrake, conducted his own experiments with Jaytee and came to quite different conclusions. Wiseman kept a continuous record of Jaytee's movements while Pamela was away, and claimed that there was too much inconsistency in the dog's behaviour to support the claim that he was responding directly to Pamela. For instance, to eliminate the possibility that the dog simply visited the window often enough that eventually he would be there when she arrived home, Wiseman decided to count as positive only trips to the window that occurred in the ten minutes following Pamela's decision to return home. However, many of the dog's movements began before the decision and so were disqualified. (Sheldrake later argued that these represented the dog's awareness of Pamela's knowledge that she would *soon* be going home.)

In further trials Wiseman, on Pamela's advice, ignored brief visits to the window on the grounds that these might not represent true awareness of her return. He instead counted only trips that lasted at least two minutes. (Sheldrake argued that this time restriction ignored trips to the window that, although lasting fewer than two minutes, were timed exactly to Pamela's trip home.) Even with this restriction, Jaytee seemed to be visiting the window at times that had been neither prompted by his mistress's return nor triggered by anything unusual outside. They were "inexplicable." Of course such inexplicable visits—random visits to the window that seem motivated by nothing in particular—reinforced Wiseman's suspicion that one such visit could accidentally coincide with Pamela's return and be mistakenly interpreted as telepathy.

Sheldrake wondered if such inexplicable incidents were really only inexplicable to us, and that the cameras or the human observers had failed to notice something outside that the dog was responding to. At one point Wiseman even agreed to redo the experiments in the winter, when there would be fewer potential distractions outdoors. Again there were, at least to his mind, no convincing results.

It would take too long to go into the detailed back-and-forth between Sheldrake and Wiseman. Every single point raised by one in one publication is rebutted by the other in another. It all came down to exactly how the respective experiments were done and evaluated. Sheldrake complained that Wiseman's experimental design unfairly excluded positive results; Wiseman argued that Sheldrake was interpreting his data incorrectly.

If you're interested in checking this out further, I'd recommend visiting the respective websites: **www.sheldrake.org/papers/Animals/** and **www.psy.herts.ac.uk/wiseman/research/papers.html**. If you do so, have your brother-in-law the statistician standing by, because it all comes down to experimental design and the conclusions you're entitled to draw from those experiments.

A Whiff
of Asparagus

Babe Ruth is remembered for his batting (and pitching) prowess, his drinking and partying capacity and even his one-liners (the best, although probably apocryphal, was delivered in 1930: when told he had made more money than President Hoover, Ruth is said to have replied, "Why not, I had a better year"), but his understanding of physiological chemistry is less well known. Yet it was the Babe who, after he had gently pushed away his asparagus salad at a posh dinner party and was asked if he didn't care for the dish, uttered this immortal line, carefully choosing his words to suit the elegance of the occasion: "Oh it's not that—it's just that asparagus makes my urine smell funny." The Babe wasn't the first to notice this effect—it apparently was first described in 1731 by the physician to Queen Anne, John Arbuthnot, who wrote, "asparagus . . . affects the urine with a fetid smell"—but he probably did more than anyone to bring it to public attention.

The curious thing is that many of the guests at that dinner party wouldn't have had a clue what the Babe was talking about, because they would never have experienced it. On the other hand, some people—and obviously the Babe was one of these—need eat only two or three asparagus spears, and within an hour their urine takes on an unforgettable, pungent odour. Some researchers liken it to the smell of boiling or rotten cabbage (although it has also been described by some, whose cooking skills are perhaps questionable, as being reminiscent of vegetable soup). Some scientists identified the chemicals that cause the odour as the sulphur-containing S-methyl thioacrylate and S-methyl-3-(methylthio)thiopropionate, both related to substances called mercaptans, key components of such delicate odours as rotting greens, onion breath and bad drains. However, the real story is practically worth a degree in chemistry.

There is no doubt that S-methyl thioacrylate and S-methyl-3-(methylthio)thiopropionate are present, but are they actually causing, or even contributing to, the odour? If they aren't volatile enough—that is, if they don't escape into the air easily enough—you will not smell them. Further chemical tests have revealed the presence of other sulphur-containing compounds, the names of which all contain some sort of combination of methyl and sulphur (dimethyl sulphide, dimethyl sulphoxide, etc.). There is circumstantial evidence that these additional compounds play a role, because they are found in smelly urine, but not in urine that lacks that essence of recently consumed asparagus.

Even that isn't the end of the story. Most of these additional sulphur compounds are small and would be easily metabolized before they ever reached the bladder (or would even be lost in the cooking process), so it is likely that they represent breakdown products of some as yet unidentified precursor.

One candidate is something called asparagusic acid, a compound that is fatal to nematode worms and may protect the plant against them. It is one sulphur compound that seems unique to asparagus, and it is more abundant when the plants are young, which of course is when we eat them. Queen Anne's Arbuthnot noted that when he wrote, "when they are older, and begin to ramify, they lose this quality."

The genetics of the ability to produce the odour, or to be sensitive to it, are as murky as the chemistry. Those people with the smelly urine apparently have a gene or genes that allow their bodies to produce these chemicals as the asparagus is digested, although it isn't known exactly how. Those without the genes never produce the chemicals and never create the odour. One problem is that the judgment of whether an odour is present or not is a subjective one, rendered by a panel of people "sensitive" to the odour. The bottom line is that there is enough variability among the studies to suggest that, as yet, there is no definitive genetic picture.

However, this much can be said: detailed family studies have shown that the gene in question can be traced back through three generations, and presumably would go back much further, although the family members that could prove that are no longer alive. Because we inherit one copy of each of our genes from both mother and father, it's possible to have a single or double dose of the "asparagus urine" gene. Parents who each have one copy of this gene will, on average, pass the trait on to three out of four of their children, but even so, it's not widespread in the general population. Studies of eight hundred medical staff and students at the

University of Birmingham showed that 43 percent had the gene, and that number agrees pretty closely with the 40 percent found in a previous English study from the 1950s. If these genetic analyses are correct, then about 6 percent of all people should have two copies of the gene, and it's interesting to note that in the Birmingham study roughly 6 percent produced urine that the testers agreed was "more pungent than the majority," something you might expect with two genes working to break down the asparagus.

While it appears that just fewer than half of us produce unusually odoriferous urine after eating asparagus, there's also a suggestion that only part of the population can smell it. The following experiment was performed at the Hebrew University in Jerusalem. A man (who incidentally had never noticed the odour) ate 450 grams of canned asparagus, and six hours later his urine was collected and divided up into a series of test tubes. Varying amounts of tap water were added to all the tubes except one. The contents then ranged in strength from undiluted to an extremely weak 1 part in 4096. Volunteers sniffed each test tube in order, starting with the most dilute, and indicated when they first smelled something.

The results were fascinating. A few people couldn't smell the traces of asparagus-breakdown products in the urine at all, even when it was undiluted. The majority could smell something odd, but only when the dilution was stronger than 1 part in 16. Finally, there was a group of smellers who stood alone in their olfactory ability. They accounted for 10 percent of the total, and they could detect the asparagus odour when it was so dilute as to be completely invisible (inolfactable?) to others. The experimenters concluded that the human species can be divided roughly into those who can and those who can't: 10 percent whose noses are extraordinarily sensitive to the chemicals produced in the urine after eating asparagus, and the rest, like the donor in this experiment, who may make the chemicals but are barely able to smell them. (And by the way, no one has any idea why we should have a gene that produces this effect in the first place.)

Of course, most of us are in the habit of urinating into a toilet bowl filled with water, which instantly dilutes the urine dramatically and masks the odour for all but the supersensitive noses. So if you can smell the after-effects of eating asparagus, you're almost certainly one of the select 10 percent who smell it most acutely. And if you produce the odour yourself, you're in the group of 40 percent of the population who do that. If you both make it and smell it, then you're allowed to brag that you and the great Babe Ruth share a common genetic heritage.

The Science
of Walking

Walking is one of those things that we do automatically, like eating or breathing. There's nothing more commonplace or everyday, but that shouldn't delude you into thinking it's simple. Walking on two legs is an extremely complicated way of getting around, and even though we adopted this method millions of years ago, we're still feeling the effects of having done it.

Each stride of normal walking involves a cascade of little tricks that we perform quite unconsciously. Are you even aware of what you do to break into stride from a standing start? When you're standing still, you're perfectly balanced, and you can't move forward unless you upset that balance. So your first movement is to relax your calf muscles, which immediately causes you to start toppling over. As you lurch forward, you quickly throw one leg forward to break your fall, and you are now walking. Once the heel of that lead leg hits the ground, you're actually in quite a stable position, one leg forward and one leg back, but only momentarily, because the rear leg pushes off, the foot rolling from back to front until, at the last moment, all the remaining weight is squarely on your big toe.*

Once your rear leg is off the ground, you swing it right through underneath your body (with surprisingly little muscular effort) and out in front, keeping the knee and ankle slightly bent so the leg doesn't hit the ground on its way through. Then that leg breaks the fall, with the heel absorbing most of the impact, and the other begins pushing off, starting the cycle all over again.

Average-sized adults can walk at most between two and a half and three metres per second, about five or six miles per hour. That speed is limited not by muscle power but by the length of the legs. Mathematical models of walking show that the longer the legs, the smaller the up-and-down movement of the torso as you walk, and the greater the maximum speed. This is why little children have to break into a run to keep up with their parents, and why people on crutches can walk surprisingly fast—their "legs" extend from their armpits to the ground. This also explains why the peculiar swaying, waddling gait used by race walkers allows them to walk about 25 percent faster than the rest of us. Their secret? They minimize the bobbing up and down of the torso, not by having longer legs, but by bending their backs and tilting their hips with each stride. That's why they can go four metres per second instead of three.

---

* If we were much heavier, the big toe couldn't handle that weight, and our pattern of walking would have to be different. Interestingly, whoever has faked (I guess I should say, *if* they're faked) the thousands of Sasquatch footprints in the Rockies has taken this into account and designed a foot that appears—from the prints—to push off with all the toes at once, something you'd have to do if you weighed as much as a Sasquatch.

You consume the least energy per metre walked if you average about 100 steps per minute—any faster and you're putting too much energy into accelerating each foot from a standing start, then stopping it as it hits the ground again. However, studies of city pedestrians show that men take about 110, and women 118 steps per minute. Obviously the pressures of the big city force you to sacrifice efficiency for speed.

The remarkable thing about all this is just how precarious walking is. You spend three-quarters of your time balanced on only one leg—more if you're walking faster—and while you do manage to move forward, you're also bobbing up and down (obvious in film scenes of oncoming pedestrians), swaying dangerously forward, and at the same time nearly toppling over sideways. You're not aware of this happening yourself, and it's not even evident when you watch someone else walk, because you're distracted by swinging arms and legs. But if you could somehow focus just on a pedestrian's torso, you'd see it rocking and rolling, up and down and side to side. It's curious that your head, even at its highest point, is never quite as high when you are walking as when you're standing still. You could walk through a tunnel that's exactly your height and never bump your head.

Even your forward motion is irregular, alternately slowing and accelerating. If you don't believe it, remember what happens when you try walking and holding a shallow pan full of water: it slops back and forth because your forward speed isn't constant. When either of your legs is out front, it's actually slowing you down; it contributes to accelerating you only when it's behind and pushing down on the ground. The thing that keeps us upright through all this up-and-down and to-and-fro motion is the subtle contraction and relaxation of the right muscles at the right time.

For instance, the moment one leg leaves the ground to swing through, your body wants to fall to the unsupported side, and it is only prevented from doing so because of the action of two members of the gluteus family of muscles. The gluteus minimus and medius on your left side contract when your right leg is off the ground, thus tightening up your left side and preventing you from falling to the right. At the same time, the pelvis rotates a little, increasing the length of the stride. Females have to rotate their pelvis slightly more to achieve a longer stride, because their hips don't have the same forward range as males'. This, of course, leads to a distinctive female walking style, of interest particularly to experts in biomechanics and manufacturers of blue jeans.

The most famous gluteus muscle of all, the gluteus maximus, not only gives the buttocks whatever shape they have but also prevents us from pitching forward by contracting and pulling us upright, especially as we climb stairs or run uphill. That tendency to jack-knife forward is much reduced on level ground, and in that situation the gluteus maximus is surprisingly unimportant—you can walk nearly normally even if it's paralyzed. The gluteus maximus is the biggest muscle in our body, and is a perfect illustration of how complicated the transition was from four legs to two. Chimps and gorillas, our closest living relatives, have gluteus maximus muscles too, but they're relatively insignificant because these animals are quadrupeds and have no need to stabilize an upright torso balanced on two legs. (Blue-jean manufacturers have no interest in chimpanzees and gorillas.)

A great many alterations are necessary to change a quadruped into a biped, and it's clear that the process began a long time ago. The oldest signs come from Africa. The skeleton of the little hominid called Lucy, found in Ethiopia, is about three million years old and already shows all the skeletal adaptations necessary for a little woman-ape to walk upright. She isn't completely modern yet—some anthropologists see in her curved toes signs that she might still have been adept at climbing—but there's no doubt she's on her way. In 1977 Mary Leakey, wife of the famed anthropologist Louis Leakey, mother of the famed anthropologist Richard Leakey, and a famed anthropologist herself, announced the discovery of two sets of footprints in volcanic ash in Tanzania that are three and three-quarter million years old and have all the features of footprints made by upright-walking creatures.

Why walking upright came at all is still an open question, but most theories these days are predicated on the idea that freeing the hands was the most important reason. There are many variations on this theme: two-legged hominids could carry their offspring and so become nomadic and exploit new food sources; or two-legged males could bring food back to a female, who, thus freed from the responsibility of gathering food herself, could spend more time having babies. But these aren't the only explanations for our two-leggedness—one recent theory even suggests that our ancestors stood up to cool off. By doing so, they would have presented less surface area to the overhead sun and absorbed 60 percent less heat.

Regardless of why our ancestors stood up, we are still paying the price for that move, at least three and three-quarter million years later. The problem is that we're walking upright using a skeleton built for four-legged travel. You don't have to go to Olduvai Gorge to dig up evidence of the shift from four legs to two: just check out your own body as you sit slumped in the living-room chair. Bothered by flat feet or fallen arches? That's what happens when you try to bear the entire weight of the body on two feet. Sore back? Four-legged creatures have no such problems, built as they are with their weight distributed evenly along the backbone. But tilt that backbone on end, and you've got discs being squeezed out of place, and a torso sitting on pelvic bones that never had to support any weight before. Suffering from a hernia? You wouldn't be if you were still on all fours, with your internal organs suspended from and supported by a horizontal backbone. Hemorrhoids? Blame those congested blood vessels on the fact that they were upended in the move from four legs to two.

On the other hand, walking is an efficient way of getting around on two legs. In 2005 two scientists at Cornell University, Manoj Srinivasan and Andy Ruina, ran a computer program to test that idea. They created a virtual human, then asked the computer to identify the most efficient ways of moving at a variety of speeds. It was a little like asking the mirror on the wall who is the fairest of them all. The computer replied that there are three ways of moving efficiently. Two were no surprise: walking is best at low speed, and running at high speed. But in between, the computer identified a mode of movement that few, if any, humans use with any regularity.

It's a blend of running and walking (and therefore could be called either "ralking" or "wunning"). It includes both the heel strike and toe push-off of walking, but incorporates those moments—unique to running—when only one foot is on the ground at any time. This curious kind of movement is efficient, but Ruina and Srinivasan say the only time you see it in action is when obese people go jogging. We have no idea whether "wunning" was used by any of our ancestors. It's unlikely given that very few hunter-gatherers are overweight.

It's funny that after millions of years of walking, we still haven't got it quite right. We sway and rock and lurch our way forward, punishing bodies that never really were designed to be upright. In that same time span our brains have trebled or even quadrupled in size, allowing us to develop language, culture and civilization. Upright walking came first (Lucy had a brain the size of a chimpanzee's, yet she was practically modern in her walking) and probably played an important role by allowing the hands to manipulate objects and work with a brain that was capable of incredible growth. Maybe walking was a kind of an off-the-shelf evolutionary solution to the problem of freeing the hands, which, once accomplished, was never fine-tuned. It's something to think about the next time you take a stroll—you probably owe your big brain to that imperfect, but unique, mode of transport.

The Swarm

It's dusk on a late-summer day, and there, hovering over the sidewalk in front of you, is a cloud of insects. This cloud, or swarm, is itself almost motionless, although the individual insects are moving ceaselessly up and down, back and forth. What on earth are they doing? The answer is simple: they are there for only one reason—mating.

Swarms are almost always bunches of males who hover together over a prominent landmark waiting for females to show up. Any female who wants to mate finds the swarm, flies up into it, is grabbed by a male, and the two drop to the ground in each other's clutches. A survey of the sexes caught by the sweep of an insect net through a swarm tells the story: numbers like 700 males and no females or 4,300 males and 22 females are typical. (Presumably those 22 were recent arrivals seeking a male.) Swarming isn't universal, but tiny insects like gnats, midges and mosquitoes, and the much bigger mayflies (fish-flies) and horse-flies all form male-only swarms. Swarms probably make successful mating more likely for both sexes: a tiny, solitary male insect aimlessly cruising around would presumably have little chance of encountering a female, whereas if he's in a swarm, females can find him. For females, going to the swarm beats scouring the bushes looking for a male. But knowing why male insects form swarms only raises more questions: how do the males choose a place to swarm? What keeps the swarm together? And if you're one of hundreds of males looking to mate, the old real-estate adage "location, location, location" must come into play: if you're waiting for a female to arrive, where's the best place in the swarm to wait?

Swarms are always hovering *over* something. It can be as simple as a white cloth on the ground or a patio stone. The males position themselves over it, facing into the wind, then fly some simple pattern that keeps them in roughly the same place. Swarms can gather over chimneys, at the tips of tree branches, over telephone poles, steel barrels, cowpies and even human beings. Frederick Knab, an American entomologist writing in 1906, described entering a field where there was a swarm of male mosquitoes over every single corn stalk, and having a swarm form immediately over his head. I saw one in Vancouver hovering over an empty Foster's Lager can. In 1807 in the German town of Neubrandenburg, an immense swarm formed above the steeple of St Mary's church, which at the time was being used as a powder magazine. The dark, columnar swarm waving gently in the wind looked so much like smoke that it convinced a number of inhabitants to head for the hills. Only when the "smoke" didn't spread did someone climb the steeple and discover the truth.

The landmark, whether it's a beer can or your head, plays a crucial role in allowing the males to hold the swarm together.

Instead of having to maintain equal spacing among themselves, as if each were a member of an aerobatic flying team, they ignore each other and pay attention instead to the landmark below them, usually maintaining their position over the upwind edge of whatever it is. You can demonstrate how out of touch they are with each other by encouraging a swarm to develop over something like a white sheet on the ground, then repeatedly cutting the sheet in half. By doing so often enough, you can reduce one original huge cloud of insects to a myriad of swarms of one—each hovering over its fragment of the original marker, apparently oblivious to the fact that it's now alone. Of course they must pay *some* attention to the others, if only to avoid collisions, and naturally they have to look out for newly arrived females, but what looks like a coordinated group activity is really just a bunch of lonely males.

But while that much is known about swarms, there's still lots of mystery. One species of midge—a tiny insect you can easily overlook until it bites you—forms not one but *two* swarms above a pad of cowdung. One is a column extending from a metre to two metres above the ground, centred on the pad. Directly below that is the second swarm, a sphere, hovering only 15 centimetres or so above the dung. Why two? The same thing happens with some mosquitoes who hover over humans: a pancake-shaped swarm gathers right over the head, and a separate column forms above that. Those males in the pancake section are thought to be waiting for females cruising in for dinner at the earlobe or the nape of the neck (in the previous example the female midges are presumably likewise attracted to the dung), but why then do the other males bother forming the column above? It would seem to be a less-desirable location—are they in the swarm equivalent of cheap seats? Nobody really knows.

Male insects live for mating, and those that fly in swarms reflect that even in the way they're built: they're engineered to detect females. Male mosquitoes are a good example.* They resemble the females in appearance, except that they have beautiful feathery "plumose" antennae sticking up from their heads. These antennae, composed of a central stalk with numerous very fine side branches, are listening devices, tuned specifically to the sound of a female; the same whining buzz that drives you nuts as you're just getting to sleep at the cottage will bring male mosquitoes from every direction.

Experiments decades ago showed you could get the same effect with tuning forks held in front of caged male mosquitoes. Their antennae vibrate with the note of the tuning fork (or the buzz of a female's wings), and tension receptors at the base record that vibration, transmitting it to the male's tiny brain. Males with amputated or immobilized antennae are deaf to the female's sound. As a female enters the swarm, the sound of her wings immediately brings the males to her. In fact you can attract a swarm of male mosquitoes yourself just by humming or singing the right note—one naturalist at an outdoor concert had the

---

* Male mosquitoes may land on your skin and walk around, but they never bite. They spend a rather quiet life supping nectar from flowers. A female, on the other hand, if given a choice of nectar or blood will *always* choose the blood. Even if their legs are amputated, females will hover over one's skin and try to pierce it. After the females have drunk their fill, their distended stomachs press on nerve fibres that in turn communicate with the brain, triggering the release of hormones that cause the female's eggs to mature.

disconcerting experience of having a swarm dive into his face every time the band hit an A.

Feathery antennae aren't the only engineering tricks that swarming males use. Blackflies and horseflies swarm too, but because they recognize females by sight, they have specially adapted eyes. A typical insect's eye is a bulging hemisphere evenly studded with individual facets, each of which produces a separate image.

But these swarming flies have a special requirement: they wait for a female to fly into the swarm above and in front of them. Instead of having two ordinary fly eyes, these males have a pair of eyes that look like aviator sunglasses: separate at the bottom but joined at the top. And the facets in the upper half are much bigger than those below. Apparently the lower half of the eye takes care of the day-to-day things in the life of these male flies, but the upper half is adapted to scan the entire airspace above and in front for the sudden appearance of a rapidly flying female. And because her tiny image is zipping across the eye from facet to facet, it's better to have fewer large ones, to give the brain time to resolve and identify the image.

Being well equipped to detect a female is only part of the battle. You not only have to find her, you have to find her first, and sometimes fractions of a second are critical. Some places in the swarm must be better than others: if, as is true in many species, females normally fly into a large swarm from below, then it's probably not a good thing to be hovering near the top. Some studies of an insect called the "lovebug," which lives in the southern United States and Mexico, did show that the swarm was segregated: big guys at the bottom, littler ones above. It's usually assumed that the big males get what they want (they are commonly the victors in the one-on-one fights for females that are often seen), so the bottom must be a better place to be. Maybe it's easier from there to scan the terrain on all sides and be ready to meet a female the instant she rises up from the ground.

It would be a mistake, however, to think that big males always have the advantage in a swarm. Nature is never so clear-cut and predictable. Athol McLachlan, a scientist at the University of Newcastle-upon-Tyne, England, has sorted through all the copulating pairs of midges caught in a net, and finds that some of the male partners in these pairs are surprisingly small. Even more amazing, these micro-males are *always* with a female. Why weren't they bullied out of the way by bigger midges? McLachlan could only guess that they must, by virtue of their small size, be great aerobats—their short wings, like short skis, giving them the ability to loop, dive and turn rings around their bigger rivals, and get to females first. The moral of this story is that swarming males must take different routes to eventual success.

Now the dark side of the story. There are always hazards involved if you're consorting in large, conspicuous groups with your rivals. Dragonflies and bats have come to recognize that a swarm containing thousands of mindless insects, however tiny, is a conveniently packaged meal. But swarming males can suffer death in another particularly gory fashion.

Some female midges enter the swarm looking not for a mate but a meal. Many of these females are no bigger than a couple of millimetres long. They fly directly into a swarm of males of another species—males sometimes their own size, sometimes much bigger—then suddenly hit the brakes and start hovering, gently moving back and forth within the male swarm. The males don't react. They all just keep doing their individual dance. Ultimately, though, a male approaches too close, to within a couple of centimetres below the female. She drops on top of him like a stone, they sink to the ground, and the female sinks her proboscis into the male's head and starts sucking him dry. These females will even fly to a likely swarm marker and hover there patiently, waiting for the unknowing males to show up. Usually they are easy pickings, because they're looking for a mate, not watching out for a dive-bombing blood-sucking female.

There's one last twist to this story of females who feast on others' swarms. Having killed several males of another species, these females then move on to swarms of their own males for mating. Unlucky males! The tipoff to their fate is that unlike most of their relatives, these males don't have feathery antennae for picking up the sound of the female as she enters the swarm. Nor do they have aviator eyes. They're blind and deaf to her arrival, but it doesn't really matter, because in this case, *she* is looking for *them*. The female simply executes the same routine she uses for feeding: she enters the swarm and grabs a male. This time, though, as the two sink to the ground, the male manages to inseminate the female while she drains him dry. When it's all over, the female throws away the dried-out husk of the male and takes off to lay her eggs. Sometimes, though, there is one telltale sign of what's happened: the female takes flight with the male's genitalia still clinging to her. And after all, for males, that's all that really counts.

THE EVOLUTION OF
THE TEDDY BEAR

I t's hard to imagine children ever growing up without a teddy bear to cuddle, but the teddy bear is actually a relative newcomer, a creature of the twentieth century. Yet a careful study of teddies shows that even in this short time, the animal has been changing shape steadily and dramatically. In a world where evolution is measured in millions of years, this is a remarkable development. The evidence for it was published in the scientific journal *Animal Behavior* by two Cambridge University biologists, Robert Hinde and L. A. Barden, who made their discovery after examining all the teddy bears in an exhibit at the Cambridge Folk Museum.

The teddy bear was named for American president Teddy Roosevelt. Roosevelt had gone to Mississippi in November 1902 to settle a boundary dispute between Louisiana and Mississippi. During his stay he did some hunting, but had no success. On the last day, someone produced a pitiful little bear, tied up, and apparently encouraged the president to fire away. Roosevelt is reported to have said, "I draw the line. If I shot that little fellow, I couldn't look my own boys in the face again." A cartoon called "Drawing the Line in Mississippi" appeared in the *Washington Post* the next day, showing the president refusing to aim his rifle at the tethered cub. The cartoon had political implications—that the president was reluctant to impose the presidential power on what was essentially a local issue—but it was Teddy's bear that became famous. Although human-like bears were already popular subjects in children's stories, this incident prompted toy manufacturers to begin making cuddly toy versions. Within five years of Roosevelt's trip to Mississippi, toy teddies were flooding the market.

According to Drs. Hinde and Barden, in those early days teddy bears looked like real bears, with low foreheads and long snouts. But that head shape has changed gradually, and the teddy bear's eyes have crept down lower and lower on the face, creating a higher and higher forehead. At the same time, the long snout has been shortening, although the authors note that this change was more or less complete by the 1930s. Today's teddy bear is an animal with a high forehead and a short muzzle, quite unlike his forebears.

One 1916 bear in this study was a particularly intriguing specimen, because it had, at that early date, the head shape and proportions of a fully modern teddy. Evolutionists can only speculate on why that particular line died out, even though it foreshadowed what was to come.

The teddy isn't the only animal to have evolved with dramatic speed over the decades. Stephen Jay Gould, the Harvard scientist and writer, noted several years ago how Mickey Mouse has changed since his introduction in the late 1920s. His eyes have become much larger, his snout thicker (giving the impression of shortness), and his ears have moved back, making his forehead look higher. These changes are similar to those seen in teddy bears, yet these are two completely unrelated species (the ancestral lines leading to bears and mice had already split when the dinosaurs were still around), with very different habits. However, they do have one outstanding trait in common: their very existence depends on their ability to charm humans.

What charms humans? The Nobel Prize–winning expert in animal behaviour, Konrad Lorenz, was the first to point out that adult humans respond to certain kinds of facial and cranial features in other humans and even animals, features that encourage nurturing and affection in the humans who see them. These include a large head, high forehead, big eyes set low in the face, chubby cheeks, short and stocky arms and clumsy movements. In other words, everything that human babies have. And everything that puppies and kittens and other baby animals have. These are the physical features that trigger the reaction "Oooooooooooohhh, isn't he sooooooo cute." A baby's survival is, of course, enhanced if its physical features elicit feelings of parental protection and love.

In a laboratory test of Lorenz's hypothesis, students at the State University of New York at Stony Brook were shown slides of a variety of line drawings of human faces and asked to rate them on a scale of one (unattractive) to seven (cute). The faces themselves differed in the location and/or size of the facial features. The eyes, nose and mouth could be near the bottom of the face, leaving a very high forehead, or at the top, producing a giant chin. Eyes varied in width and height, and even in the diameter of the iris. Only one feature was changed in every slide, so there was a bewildering array of faces, but even so, one face stood out in the ratings. It was a baby-like face with all the features that Lorenz had predicted: large eyes set low in the face, exposing a high forehead.

Both the teddy bear and Mickey Mouse have been evolving to secure their place in the human heart by tending more and more to infantile features. This process, well known in zoological circles, is called neoteny. Any species that is evolving in this manner prolongs its juvenile stage and so delays adulthood. The result is that the adults of today look like the young of past generations.* Neoteny is seen in organisms as diverse as salamanders, some of which reach sexual maturity without ever losing their gills, and flowers. One species of larkspur has slowed its development so much that the mature flower looks more like the buds of other species. This juvenile tube-like flower is suited to pollination by hummingbirds, a lucky break for both plant and bird.

As Hinde and Barden point out, there are other interesting physical changes seen over the years in toy bears. One particular specimen, which first appeared in the 1920s, was striking because of its tiny brain case—measurements revealed a very short distance between its eyes and the top of its head. These anatomical data support the unusually well-documented behaviour of this bear, as copious field notes attest to the animal's apparent lack of intelligence. He is known as Winnie-the-Pooh.

---

* In fact, humans have evolved by that very process. We have become human-like rather than ape-like by slowing down our development and retaining fetal or child-like physical characteristics into adulthood. An adult human looks much more like a fetal chimpanzee than an adult. We also presumably look much more like a fetal *Homo erectus* than an adult. You could easily speculate that if this trend continues, future human adults will have even flatter faces and bigger heads relative to their bodies, like modern human fetuses.

This study is an important step in paleoanthropology. Hinde and Barden have been able to trace the evolution of the teddy in detail from its initial appearance in the fossil record to the modern form. Rarely is such a complete record available. Their data confirm the Lorenzian speculation that evolution will favour high foreheads and short muzzles to foster parental nurturing behaviour. And most important, any hitherto unclassified teddies discovered in cottage or attic excavations can be placed in their appropriate position in the teddy bear family tree by a simple measurement of their cranial dimensions.

I'll Never Forget
What's-His-Face . . .

Picture a face you know really well: someone in your family, a friend or even someone famous. Exactly what is it about that face that you remember? Is it a particular feature like the nose, forehead or mouth, or do you remember the face as a whole? Psychologists today would be hard-pressed to say which of these two different strategies—or any other—we use to recognize faces, even though we're doing it all the time in our everyday lives.

Even newborn infants a few *minutes* old will spend more time looking at and following the movements of a face-like pattern than any other, even if it's as simple as two small squares for eyes and another small square for a nose. This surely shows how fundamental face recognition is to humans. If you mix up those three simple features—put the nose on the side and the eyes one over the other—the infant loses interest. This early recognition ability isn't very sophisticated, because if the "face" stops moving, the baby quickly loses interest, but by two to three months, babies have a better system in place, and will prefer a proper face to a scrambled one, even if it isn't moving.

As toddlers become children, they get better at remembering faces, but even six- to eight-year-olds are easily thrown off by disguises like wigs or glasses. (In fact, some studies have shown that adults, too, won't remember faces they've seen briefly if those faces were wearing glasses.) If their classmates wear hats and glasses, children aren't fooled, but when the faces are unfamiliar, they pay attention to the paraphernalia and not to the face itself. Children of this age do about as well on tests of face recognition as adults who have suffered some damage to the right hemispheres of their brains—the area where much of facial recognition goes on.

One of the strangest observations is that around adolescence, somewhere between the ages of eleven and fourteen, the ability to recognize faces suddenly slumps, and for a couple of years either gets worse or at best levels out before beginning to improve again. It's not yet clear why this happens. Another curious finding is that while we get better at recognizing normal faces as we mature, even as adults we're no better than preschoolers at recognizing upside-down faces. One possible explanation is that by the time we're adults, we've locked in the brain routines for recognizing right-side-up faces, and an upside-down face simply doesn't compute. Preschoolers, on the other hand, haven't perfected their ability to recognize any face yet, and so an upside-down face isn't as much of a foreign object to them as it is to an adult.

Just how well we remember faces was revealed in a study done in the mid-seventies, in which people were shown five faces from yearbooks dating back to the time they had been in high school, and were asked to pick out the one person who had been a school-mate. Even those who'd been out of school for thirty-five years were able to recognize their schoolmate 90 percent of the time, an amazing figure (and one that shows that all the new faces they had learned in the interim hadn't interfered with the original memories). The success rate dropped to around 75 percent for people who had been out of school for forty years. Remarkably, fifty years had to pass after graduation before the ability to recognize classmates' faces started to drop significantly. Nor did it matter how big the school was: one hundred or eight hundred students, the results were about the same.

This study also showed how much easier it is to recognize a face from the past than to recall a name. Even recent graduates (three months out of school) were able to list by name an average of only 15 percent of fellow students, regardless of the number they'd gone to school with. That shrunk to 9 percent for those who'd been out of school for forty years. And how about this curious observation: both men and women remembered twice as many boys' names as girls'!

The main problem in face-recognition research is determining what attracts our attention. Do some features lend themselves to recognition better than others? A typical example of the way psychologists try to address this question was a British study in which forty subjects wrote descriptions of ten faces while they were looking at black-and-white photographs of them. The experimenters (or more likely their graduate students) went through all four hundred descriptions, tabulating how often each part of the face was mentioned. Then a second experiment was done, where colour photographs of only two faces were used. Subjects were asked to describe one face immediately after seeing it, and the other after a delay of an hour, a day or a week. While these two experiments were quite different, the results were almost exactly the same: hair was the feature quoted most often, followed by eyes, nose, face structure and eyebrows (note these are all in the upper half of the face), then chin, lips, mouth, complexion and cheeks. If this order indeed represents degrees of importance in recognition, then the Lone Ranger was really smart to wear a hat and a mask, thus hiding his hair, eyes, eyebrows and most of his nose—four of the first five on the list. The last, and presumably least important, feature was the forehead, suggesting that trying to disguise yourself by wearing a headband is not a smart thing to do.

The problem is that for every experiment that comes up with a list of facial features in one order of importance, there's another that contradicts it. For instance, studies that present isolated parts of faces to subjects to see if they can recognize the whole from the part ("face fragmentation" studies) rank the eyes most important, followed by mouth, then nose. Others that change one or more features to see if those changes affect recognizability ("face distortion" studies) suggest that hair is most important, followed by eyes and chin. Some studies even suggest that specific facial features like eyes and nose are not as crucial for recognition as general qualities, like the age and overall shape of the face.

Obviously these studies, however well designed they are, can give at best only very rough indications of what goes on in our brains when we see a real face, remember it, then recognize it later. All the above tests use pictures, but in everyday life we see live faces that move, express emotion, cause emotional reactions in us and take on dozens (hundreds?) of different looks in only a few minutes. Questionnaires and tests can only take you so far. But some further hints of the complexities of face recognition have been revealed by the discovery of people who can't do it.

They suffer from a rare condition called prosopagnosia—the inability to recognize faces. A face to them is an unidentifiable object, whether it's a family member's face or even their own. (In one famous case in the nineteenth century a patient started to excuse himself to a man blocking his path in a hallway, only to realize it was his own reflection in the mirror.) Many prosopagnosics are aware that it is a face in front of them, and that there's a nose, two eyes, a chin and so on, but they can't put it all together in a recognizable form. Some people with this problem look at a face and see only a vagueness that can't be recognized, while others see the features jumbled or disarranged, as if they're viewing some living Cubist painting.

Prosopagnosics are forced to recognize people by their tone of voice, clothing, mannerisms of movement, sometimes even by a beard, moustache or hairstyle. Show a prosopagnosic a famous face, like Michael Jackson's, and he or she might be able to say that he's a music star, that he was on television last week, even that he was singing and dancing, but still be unable to go any further. Having been told it's Michael Jackson, the person might agree, understand, then be unable to recognize his face again the next day.

The inability to recognize faces seems to result from damage to areas near the back of the brain. At first it was thought this was a disability of the right cerebral hemisphere, the right side of the "thinking" brain. This made sense because in most people the right hemisphere is particularly good at spatial organization, putting together objects like noses and ears into a recognizable whole. On the other hand, the left hemisphere, skilled in assembling sequential information like language, might be content with a simple list of facial features. However, it now seems that in most cases of prosopagnosia there are problems on both sides of the brain, and that each hemisphere plays a role in face recognition. The right may be good at long-distance identification—knowing that the object on top of that torso forty metres away is a face—with the left taking over when fine close-up detailed analysis is necessary.

Recent research has suggested that these unfortunate people actually do recognize faces—at least their brains are aware that they're seeing a familiar face—but that awareness never enters the person's consciousness. Antonio Damasio of the University of Iowa recorded changes in the electrical conductivity of the skin of prosopagnosics looking at pictures of both people they knew and people they didn't. These conductivity changes indicate some nervous or emotional reaction (they're the basis of the lie-detector test) and in this instance appeared only when the patients were looking at the faces of people they knew. Yet they weren't aware they were looking at familiar faces. Damasio speculates that we store the memory for a face in one place in the brain and the memories associated with that face somewhere else, and recognition requires both the vision of the face and the memories that go with it. He claims that these people can't bring them together properly.

There have been some dramatic advances in the understanding of what parts of the brain are responsible for the recognition of faces. A lot of attention has been focused on an area just behind the ear called the fusiform face area, or FFA. Magnetic resonance imaging has shown that this small patch of brain tissue responds most actively to images of faces. It doesn't act on its own, especially when it comes to recognition (other areas kick in then), but it is the foremost face module. There has been controversy over the FFA. Some researchers have argued that it is not specific to faces but is rather a really good processor of details about familiar objects. So, that argument went, bird watchers or car experts would activate their FFA to determine exactly what species or model they were looking at. But the most recent experiments suggest that the FFA is at its best when looking at faces (although it is true that there have been cases of prosopagnosia, not for faces, but for birds, and even for cows!).

But of course the FFA isn't all there is to it. At least two other main areas of the brain, nearby, participate in the perception of faces. These companion areas likely play a role in analyzing movements of the face, particularly the lips when speaking, and facial expressions. In turn these areas link to others, such as parts of the brain that read the emotional content of facial expressions, and those that remember the salient facts about that face, if it is a familiar one. One group of researchers divides these face areas roughly into two kinds: one set performs the role of identification and recognition, the other analyzes the movements and expressions that are so important for social communication. Identify the face first, recognize it, then react to it.

Face research has even substantiated some folk beliefs about faces. For instance, the old saw "they all look the same to me" turns out to be true, no matter which group—blacks, whites, Orientals or South Asians—has been studied. People all have more difficulty recognizing faces belonging to a group other than their own. One suggested explanation for this inability is that any person will concentrate on features that are uncommonly seen in his or her own group: Caucasians will concentrate on Orientals' eyes, Orientals on Caucasians' noses. If that person is then asked to pick out the face that he has seen from a series of like faces, he'll perform poorly. There's some research that suggests people who spend a lot of time with other groups are better at recognizing individuals within these groups. Studies have even been done to make sure that there is no group in which faces are truly "all the same," more uniform than other faces. There are no such groups.

And finally, something you might have suspected all along: beautiful faces are remembered more easily. And we don't need long to register a face as being attractive: a 150-millisecond glance—less than two-tenths of a second—gives the same ratings of attractiveness as unlimited perusal. Beauty is a rare thing, too, if you believe the judgments of experimental subjects, because when asked to classify faces according to attractiveness, they'll throw many into the lowest rating (the least attractive), but will almost never classify a face in the highest or even second-highest rating. It's as if we rate faces in the same way that judges rate skaters or gymnasts—never award too high a mark to the one you're looking at, in case a better one comes up. Hope springs eternal, even in the unconscious brain.

An Uplifting
Experience

If you live on the prairies, one of the signposts of spring and fall is the occasional V of geese flying overhead. That V is a striking and rare formation among birds, but it's also the subject of controversy, because scientists can't agree on why geese should bother flying in Vs. The debate centres largely around the claim that geese arrange themselves in Vs to be able to take a ride on upwelling air coming from the wingtips of other geese. A branch of mathematics developed to analyze the aerodynamics of aircraft has been applied to geese, and the numbers leave no doubt that flying together is the best way to fly: one set of calculations showed that a formation of twenty-five birds, tip to tip, could increase their range by 71 percent over single birds expending the same amount of energy. Nine birds could gain about 50 percent, and even three birds would derive some benefit.

Those numbers actually apply to flying abreast in a line, but there's a mathematical expression called Munk's stagger theorem that says you can rearrange a line of flying objects into a V and still provide an uplifting experience, the only difference being that some geese will lose benefits and others will gain. The lead bird loses some lift because his nearest neighbours are now behind him (although surprisingly he's still better off than he would be on his own), but the birds at the tip, while they're only getting lift from one side, are also receiving the benefits of upwash drifting back from birds farther ahead in the V.

Here's how the aerodynamics should work: as it flies, each goose creates turbulence and eddies of air swirling away from its wings. The air immediately behind the wings is moving down (it's called the downwash), but to the sides, just beyond the wingtip, it's actually moving up, and at a pretty good speed too. The intensity of this upwash falls off quickly, though, so that if you were as much as a metre away, it wouldn't amount to much. If you're a goose, you shouldn't fly directly behind another, because you'll be fighting the downwash; instead, flank that goose and rise just outside the wingtip, and you'll get lift from the upwash.

The theory provides a few other interesting details: the best V would actually be a little bent, with the tip birds drifting slightly back and out from the main line, and the lead bird slightly back toward the rest of the pack. The angle of the V should be roughly 100 degrees, about the angle your little finger makes with your thumb if you stretch your hand as wide as possible, but it does not have to be symmetrical. A V with ten birds in one arm and fifteen in the other would require only slight adjustments in spacing to work. The really neat thing is that it's easier to maintain the V shape than it is to disrupt it. If a goose flies too far forward out of the V, he'll lose uplift and fall back into line. If he drifts too far back, he'll get more lift, and suddenly find himself surging forward back into his proper place. (It would be possible, of course, to be a complete parasite by falling back and at the same time flapping less vigorously. The little vortices of upwelling air in the interior of the V would permit a goose there to burn fewer calories, and so reap the benefits without contributing to the good of the whole. There must be some strong social pressure against doing this.)

If Vs are so great, why don't more birds use them? Why don't we see Vs of sparrows or robins? Again, the aerodynamic analysis provides the answer: the size of the upwash behind a wing depends on the size of the wing itself. The smaller the bird, the smaller the wake, and the closer they'd have to fly to each other to benefit from it. You'd have to have a raft of sparrows, not a V.

Of course, in science it isn't enough to have a good idea— you have to prove it. Proving this one has been a lot tougher than you might have thought. It would be nice if you could persuade eight or nine Canada geese to take to the air simultaneously and hold a perfect V in a wind tunnel just long enough to be able to visualize the air movements around the wings. This experiment is still waiting for some dedicated researcher. Meanwhile, goose-watchers have taken to the fields to photograph the actual Vs as they fly overhead. To their chagrin, none of the Vs that have been photographed had the exact shape predicted by the theory. But no one is throwing out the theory just yet, because it's very tricky measuring the angle of a V of geese. They approach a camera from all different directions and distances—almost never directly overhead—and even if the geese pass directly in front of you, as if in review, there's a perception problem. The angle of the V will appear to change constantly as the flock moves from left to right (or vice versa), first shrinking, then growing again. The scientists doing these studies have spent most of their time figuring out how to interpret the results they have.

Then—as always in science—there are those who want to go back and reinterpret some of the data that have already been gathered. Some researchers are now downplaying the importance of the V angle and suggesting instead that the important thing is the position of the wingtips. Each goose, they say, should be positioned so that its wingtips cross a line drawn straight back from the wingtips of the goose just ahead. That is, if they were flying abreast, their wingtips would overlap, actually brushing against each other, although in the staggered formation of the V there's no such contact. Armed with this new idea, these researchers have gone back and re-examined already published photos—those same photos that are causing other researchers so much consternation. They claim that the geese in those flocks were indeed maintaining positions like this.

While the debate rages over whether it's the angle of the V or the alignment of the wingtips that's most important, there are still awkward observations that have to be explained. If V flying is so energy efficient, why are geese apparently quite happy to abandon Vs and show up in echelons (staggered single files), clusters, Js, inverted Js, you name it? And why, even when they are in Vs, do the birds change position so much, drifting in and out, back and forth, even switching positions? Shouldn't there be one optimum position for every goose, even allowing for a little bit of adjusting to gusts of wind?*

Could it be something other than aerodynamics that prompts geese to fly in Vs? One suggestion is that geese fly in Vs to provide the best view for every bird. Flying in a straight line would deprive those near the end the sight of the leading birds. Flying abreast would of course mean there *was* no leader. In a slightly macabre twist, researchers examined the severed heads of five Canada geese killed by hunters to determine their visual capabilities, and found that the geese have a blind spot about sixty degrees across at the back of the head. Any variant of the V, like the echelon or the J, would ensure that any goose would be able to see the geese behind him or her, because they would be outside that blind spot. However, observations of real-life geese suggest that they sometimes fly in formations that would put some of those near the back in the blind spot of those ahead, so the blind-spot argument is at best only part of the answer.

There are more suggestions. One is that flying in a V provides a highly visible "advertisement" to other geese, prompting them to follow, creating larger and larger groups of birds that then have a greater communal knowledge of where to get food. If they didn't fly in instantly recognizable patterns, this theory suggests, goose

---

* Geese may still represent an unsolved problem, but a group of French researchers showed in 2001 that pelicans in Africa definitely do benefit from the uplift provided by a V. They trained white pelicans in a national park in Senegal to fly after a motor boat (or an ultralight airplane) while their heart rate was recorded. As long as the birds were flying in V formation, their heart rate was more than 10 percent slower than otherwise, and their wing-beat frequency was reduced.

populations would remain scattered. How you would prove this one is beyond me.

In 2004 two Swedish scientists, Malte Andersson and Johan Wallander, threw another idea into the mix. They suggested that flying in Vs might be a family affair. They argued that because the lead goose really derives little if any aerodynamic advantage, there must be some other reason for volunteering to fly in that position. If, Andersson and Wallander suggest, a small group of birds consists of parents and siblings, then any lead bird, by establishing the V and thus making flight easier for the others, is benefiting in a genetic sense. If those other birds survive the flight to have more offspring, and the offspring are related to the lead bird, then that bird has succeeded in the evolutionary sense of passing its genes on to the next generation, albeit indirectly.

They even make the prediction (which unfortunately requires yet more observations of the shape of Vs) that if birds are related, they would be willing to fly in an acutely angled V, where the lead birds gets almost no aerodynamic advantage. But if birds are unrelated, no leader would volunteer, and such Vs should be much more strung out, obtusely angled, a formation where the disadvantage to the lead bird is greatly reduced.

We might have to wait until the next century before any of the explanations for geese flying in Vs is generally accepted. But next time you're out on a typical Canadian fall day and you hear them honking, look up: if you're lucky enough to have the geese flying directly over you, check out the angle of the V and the alignment of those wingtips. Does it look as though they're using each other's updrafts? Are you one of those who suspect there is something else going on? The one thing you can be sure of is that geese have some good reason for flying in Vs—we just haven't figured it out yet.

When the Lavatory
Becomes a Laboratory

Scientists who study behaviour are always on the lookout for the ideal place to observe their subjects, whether it's a crowded waterhole on the Serengeti plain or a treadmill in a cage. It is to the scientist's advantage to find a location where animals perform obvious and repetitive behaviours. One such place is a public washroom. Psychologists can take advantage of the fact that public washrooms are an ideal setting to witness human behaviour free of the complications present in large social groups.

Sometimes the investigators do not even need to enter the washroom. A researcher for the Washington State Department of Transportation, stopwatch in hand, watched people enter roadside washrooms and found that, on average, men spend forty-five seconds in the washroom, women seventy-nine. A man's first reaction is to say, "Oh, I knew that women took longer anyway." Washroom research is like that: it reveals things you're pretty sure you knew, but it goes further by attaching some numbers to the phenomena.

However, while it's interesting to know that men spend, on average, thirty-four seconds fewer per visit in the washroom than women, it's a bit like watching woodpeckers vanish into their holes and then emerge a short time later: you want to know what went on inside. Woodpecker researchers could probably resort to using a hidden camera, but washroom researchers usually have to enter the research site itself to record their observations. However, they must be careful to maintain their role as observers and not confound the experiment by becoming part of the scene. This can create odd interactions. Take this exchange between a visitor to a men's washroom and a researcher who happens to be an acquaintance of the visitor. The researcher is busy writing an account of a previous visitor's actions, and trying desperately not to become involved in conversation:

Visitor: "Hey! [walks to a urinal] Nothing like pissin'."
Researcher: "Yup."
V: What th' hell ya doin'?" [walks over to the sink]
R: "Writing."
V: "Heh heh, yeah. About people pissin'?"
R: "Yup."
V: "Take care."
R: "Uh-huh."

Some novelists would be proud of that minimalist piece of dialogue, but the social psychologists interested in public-washroom behaviour would immediately recognize that as a typical encounter: there's always a conflict in the washroom between acknowledging friendship and respecting privacy—saying something or saying nothing. For instance, it has been observed that two men who know each other will stand at adjacent urinals and hold a normal conversation, but never look at each other. They maintain their relationship while honouring the other's right to privacy. Two strangers standing at adjacent urinals won't even acknowledge that there's another living thing in the vicinity. The walls of toilet stalls act as even stronger barriers to conversation: how rare it is to hear someone inside a stall talking to anyone else in the washroom. I'm reminded of the old Edmonton Gardens, where I used to go occasionally to see hockey games or wrestling. The men's washroom there featured a huge circular urinal, with men standing around it, packed shoulder to shoulder, creating a reverse fountain. I can't remember hearing any conversations.

It's possible for clever observers to take advantage of the fact that they influence washroom behaviour by their very presence. In 1986 the journal *Perceptual and Motor Skills* published a paper entitled "Effects of an Observer on Conformity to Handwashing Norm." The title means "Do People Wash Their Hands More Often If They Think There's Someone Watching?" The hypothesis was that women using public washrooms would wash their hands if there were other women around but wouldn't bother if they were alone. The study was conducted on "the campus of a large privately owned university in the western part of the United States." In the first part of the experiment, the observer was in a lounge adjoining the washroom, clearly visible to the subjects when they left their stalls. Of the twenty women who therefore saw the observer in the lounge, eighteen washed their hands. In the second part, the observer hid in one of the stalls, with her feet up and an Out of Order sign on the door. Of the nineteen women who then used the washroom, only three washed their hands.

This study seems to demonstrate that social pressure is a much more important determinant of hand-washing behaviour than concern about personal hygiene. (A majority of people I've talked to claim not to be surprised in the least by the results of this experiment, while maintaining that *they* always wash their hands.)

Since the SARS scare, it has been my casual observation that people are more conscious of washing their hands in the public washroom, but it's also true that most of those people really don't know where the threat lurks. For instance, there are those who carefully lay down strips of toilet paper on the toilet seat before sitting down. Why? The toilet is one of the least likely sources of infection in the entire public washroom. You're more likely to inhale bacteria- or virus-laden droplets of water thrown into the air from the flush of the toilet—with the risk that implies—than you are to get infected on the skin of your butt from the toilet seat. This is at least partly because bacteria adhere to the porcelain walls of the toilet bowl between flushes, reproducing happily, and are thrown into the air by the next flush, tens of thousands each time. They don't just fly into the air and then fall back into the bowl— studies have shown that the bacteria launched from the toilet end up on surfaces all over the washroom. The countertops around the sinks, the handles on the paper-towel holders, even the washroom door handle are likely more contaminated than the toilet seat.

In fact the toilet seat represents a lesser risk than the kitchen! One study surveyed fourteen household sites, including kitchen faucet handles, drain, countertops, dishcloth and cutting boards, as well as the old bathroom standbys of sink drains, countertops, the floor near the toilet, the toilet flush handle and, of course, the toilet seat. The sites most highly contaminated with fecal coliform bacteria were the kitchen faucet handles, the dishcloth and the sink drains in both kitchen and bathroom. The bathroom countertop and the toilet seat were the least contaminated of all. Maybe we should be tying strips of toilet paper around the faucets in the kitchen!

The irony is that, given that we infect ourselves often by touching our contaminated hands to our faces and noses, those poor deluded people who apply the toilet paper to the seat and inadvertently touch the seat with their fingers might be doing more to ensure they become infected than if they'd just sat down promptly.

Psychologists again turned to public washrooms to study the effect on a man when his personal space is violated. All kinds of experiments have been conducted to try to define exactly what personal space is, how big it is, and most important, how we react if it's invaded. The problem in this latter instance has always been to find an experimental situation where someone's stress can unambiguously be traced to a violation of his space.

There's one particular setting in which a man's personal space is always invaded: when he's forced to stand next to a stranger at the urinal. The stress in this situation could be measurable, the psychologists theorized, because research has confirmed that fear and anxiety can produce an inability to urinate. A fearful urinator is the victim of powerful opposing physiological forces: the bladder muscles contract, forcing urine into the urethra, but the sphincter muscles that must relax to let the urine out are paralyzed by fear. So if having someone stand at an adjacent urinal does invade your personal space, the stress that results might make it harder for you to start urinating and harder to continue, too. ("Delay in onset and decrease in the persistence of micturation," as it is phrased in psychological jargon.)

The experiment to determine if personal-space violation would actually cause men to stand and wait was conducted in a lavatory at a midwestern U.S. university (where all such experiments seem to be conducted). It contained two toilet stalls and three urinals. The unsuspecting experimental subjects were faced with three possible situations: they were alone, the other two urinals "out of commission" with Don't Use signs on them; or there was one other man (a confederate of the experimenter) present, separated from the subject by an "out of commission" urinal; or there was a man next to the subject, with the Don't Use sign occupying the end urinal. So subjects experienced either no threat to their personal space, a minor trespass or a major invasion.

For this experiment to work, it was necessary to time the subjects' urination—both the beginning and the duration. But it was soon discovered that listening for splashing, although it had worked in a pilot project, wasn't going to this time. These urinals were just too quiet. So the observer sat hidden in one of the stalls with a pile of books at his feet. Tucked in the pile was a periscope, positioned so that it was aimed up through the space below the stall wall and onto the "lower torso" of the urinal user (but not, the authors of the study hastened to say, his face). Once in position, the observer, like an ornithologist concealed by a duck-blind, waited for the subjects to come to him.

A student coming into the washroom found only one urinal available. As he stepped up to use it, the observer, watching intently through his periscope, started two stopwatches: one he stopped when the urine started to flow, the other when it stopped. So he knew how long it took to begin, and how long it lasted.

And the results? When the subject had the urinals to himself, delay in the onset of urination was, on average, 4.9 seconds, but the presence of a stranger one urinal away increased that time to 6.2 seconds. When the two men were shoulder to shoulder, the delay lengthened to 8.4 seconds. The results were, as they say, statistically significant.

To the experimenters' credit, their predictions were borne out, although not without some controversy. As you might expect, the psychologists who designed this experiment were accused of stretching the ethics a little by training periscopes on the groins of unsuspecting experimental subjects.

In 1987 psychologists brought one of the "hot" topics in animal behaviour into the public washroom. Odours as messages are routinely used in the animal world, whether it's your dog sniffing the tree to see who's been there last or a male moth flying hundreds of metres upwind to find the female who's emitting her scent. One of the big unanswered questions is: how important are these odoriferous molecules, technically called pheromones, in human communication? Not perfumes or aftershave, of course, but more subtle, naturally produced odours.

There are some hints that we humans do indeed produce and respond to these odours. One chemical, called androstenol, which male pigs emit to persuade females to mate, is also found in human armpit sweat and urine. Of course, finding it there doesn't prove that we actually respond to it, and experiments designed to prove that we do have produced mixed results. But this murky picture changed in 1987 when a group at the University of Southern California had the bright idea of testing androstenol in a public washroom: they wanted to know if it might send the same message as the chemicals that dogs spray on fire hydrants—the "I have been here" declaration of territoriality.

During an experiment that ran for five weeks, researchers taped two-inch by two-inch Plexiglas squares, at about eye level, to each of the doors of the four stalls in the men's washroom. On weeks one, three and five, common lab alcohol was dribbled onto these sheets. The presence of a faint odour of alcohol had no effect on the pattern of usage of the stalls. But on week two, a solution of androstenol in alcohol was deposited on the square on one of the stalls. The use of that stall dropped dramatically, from around twenty visits per week to a mere four when androstenol was present. (In fact, those four visits might all have been made by the same person.)

These results can't be explained simply by assuming that the men smelled the androstenol coming from the stall and chose another. It does not have an overwhelmingly powerful odour, nor is it easy to associate with urine or sweat. It's musky but faint. Andrew Gustavson, the scientist who designed the experiment, is trained to recognize the odour of androstenol, but even he couldn't detect it in this washroom until he was halfway into the room. So how could this puny odour have somehow directed men away from the treated stall?

Further evidence that the deterrence was not caused by the presence of odour alone was provided in week number four. The Plexiglas square on the same stall was soaked this time not with androstenol but with the closely related chemical androsterone. Even though it has the same sort of musky odour, it had no effect on stall choice. Only when androstenol was present on the Plexiglas square did most men avoid using that stall, as if it were exerting a territorial "don't come near me" effect. (And it is a male phenomenon: a duplicate experiment run in the women's washroom had no effect whatsoever.)

The experimenters reported that the men who avoided the androstenol stall did so in a very natural way. There was no walking right up to the test stall, pausing, sniffing, then backing off to go into another one. There weren't even sudden changes of direction. People just came in and headed smoothly for a stall—but never the one with the androstenol.

Did any thoughts go through their conscious minds as they entered the washroom? Were they even aware that they'd chosen a stall without androstenol?

Unfortunately we'll never know whether these men realized that anything unusual was happening. The ethics committee overseeing this experiment forbade the experimenters from quizzing the washroom users about their thoughts and feelings. However, it would not be surprising if they weren't aware of anything unusual. Airborne androstenol could simply enter the olfactory nerve and exert its behavioural effects via nerve circuits that would not require thinking or considering. Whatever went on in the subjects' minds, it's clear that something very odd happened during the second week of this experiment.

Would you have believed that men coming into a washroom would unhesitatingly avoid a particular stall because of a barely perceptible odour emanating from the stall door? If these results can be substantiated, they'll go into the books as the first really solid evidence that we alter our behaviour in response to extremely subtle odours. Unfortunately, in the years since this study was performed, to my knowledge there have been no attempts to replicate it. If that is ever done and the same results are obtained, it will prove once again that as a setting for studying humans and the psychology of their everyday lives, there's no place like a public washroom.

Closet Science

If you have a package of wintergreen Lifesavers and a mirror, wait until no one's looking, then sneak into the nearest dark closet. There, wait for five minutes to allow your eyes to become adapted to the dark, then pop a couple of the candies into your mouth and crunch them. Look at your mouth in the mirror as you're biting down on the candies, and you'll see flashes of blue-green light. I know it sounds unbelievable, but the flashes can be so bright that you may not even need to close the door completely; if it's really dark, the effect is spectacular.

In fact you can get roughly the same effect by crushing lumps of sugar with pliers, but you'd probably find that the light is much dimmer and a slightly different colour, more blue and less green. Robert Boyle, one of the great scientists of the seventeenth century, besides spending much of his time experimenting with air and vacuums (he showed that nothing would burn in a vacuum and animals couldn't live in one), also fooled around with sugar and noted that "hard sugar being nimbly scraped with a knife would afford a sparkling light." So although the discovery of the phenomenon is by no means new (in fact, Francis Bacon noticed it even before Boyle), the explanation of this unexpected light had to wait for twentieth-century chemistry.

Sugar is usually in a crystalline form, and while you probably know what such crystals look like as they sit in your spoon, it's how they're put together at the atomic level that's important here. A perfect crystal is mostly empty space—a skeletal, three-dimensional structure of atoms, anywhere from hundreds to trillions of them, bonded rigidly to each other. As long as the crystal doesn't get so hot that the atoms vibrate too vigorously and separate from each other, or as long as there aren't water molecules to nibble at its outer edges and begin to disrupt its structure, the atoms in a crystal will stay locked to each other.

But bring a sugar crystal in contact with two back teeth, and chaos ensues. Neighbouring atoms are torn apart, creating cracks in the crystal, and as the cracks spread, little islands of negative and positive electrical charge are left isolated on separating fragments of the crystal. When the distance between them gets to be too much to bear (chemically, of course), the negatively charged particles—electrons—leap across the gap to neutralize the positive charges on the other side. As the electrons fly through the air, they collide with molecules of nitrogen gas, the most common gas in the air. Each nitrogen molecule absorbs the electrons' energy, then releases it immediately in the form of blue light, producing a microscopic version of lightning. It's easy to demonstrate in the lab that it's nitrogen from the air that's kicking out its newly acquired energy as blue light: if you bash some sugar in a vacuum (wouldn't Robert Boyle have been pleased?), where there's no nitrogen gas present, no light appears.

We can see some of the light produced from the sugar in the candy, but much of it is in the ultraviolet range of the spectrum, meaning that although you might be able to get a very faint tan from standing in front of someone who's crunching tons of sugar, you can't see a lot of the light released. That's where the wintergreen comes in: it just so happens that oil of wintergreen absorbs ultraviolet light very well, and when it's excited by the extra energy provided by that light, it re-emits visible blue-green light. So when there are sugar and wintergreen together, the amount of light you can see is much greater, and it's a different colour than the light from sugar alone.

The same principle produces the brilliance of the pens you use for highlighting text—the ink absorbs UV light and re-emits it in the visible range, adding brilliance to the stroke of the pen. But crunching candies is so much more dramatic. And so much tastier.

LETTERS
FROM HEAVEN

In February 1635, the philosopher and scientist René Descartes was in Amsterdam for a couple of snowy days, and his descriptions of the snowflakes he saw have never been bettered:

> What astonished me most was that among the grains which fell last I noticed some which had around them six little teeth, like clockmaker's wheels . . .

> On the morning of the following day, [there were] little plates of ice, very flat, very polished, very transparent, but so perfectly formed in hexagons that it is impossible for men to make anything so exact . . .

> There followed after this little crystal columns, decorated at each end with a six-petalled rose a little larger than their base.

Descartes' brief descriptions go to the heart of what's both so obvious and so subtle about snowflakes. Every kid knows that snowflakes are six-sided. But scientists still haven't solved the mystery of how all snowflakes can be six-sided and yet take so many different shapes. If snowflakes are all made in roughly the same way, how is it that some are like "clockmaker's wheels" and others "crystal columns," miniature six-sided pillars? (Incidentally, Descartes was fortunate to see those columns capped on either end by a "six-petalled rose," like wheels attached to an axle—many veteran snowflake watchers see these only once or twice in a lifetime.)

What's amazing about Descartes' description is that it's apparently only the second time anyone in Europe acknowledged, in print, that all snowflakes are six-sided! The credit for first mention goes to Johannes Kepler, an astronomer known mostly for his work on the shapes of planetary orbits, who in 1611 wrote a little book called *The Six-Cornered Snowflake.* If you go back a half-century before that, it's apparent that the hexagonal shape of snowflakes was not common knowledge in Europe. A woodcut in a 1555 book by the Archbishop of Uppsala shows twenty-three versions of snowflakes, only one of which is six-sided. The variety of the other shapes depicted—eyes, hands, bells, arrows and half-moons—makes it clear that the author had no idea which shape was correct. In contrast, the Chinese had described the true shape of snowflakes at least as early as the second century AD.

At first glance it seems ridiculous that four hundred years ago no one in Europe knew, or at least publicly acknowledged, that all snowflakes are six-sided. You usually don't have to spend more than a few moments in falling snow before you can find individual six-sided flakes on the windshield of your car or the sleeve of your coat. On the other hand, it's always easier to see something if you already know it exists—would we notice those perfect little hexagons so readily if we hadn't been taught from kindergarten that they're there?

Recognizing that snowflakes are six-sided is only the beginning—even Kepler realized that the really interesting question is why. What mould or force could produce millions of flakes out of frozen water that, whether flat plates or columns or stars, all have six sides? He figured that snowflakes must be assembled from smaller sub-units, and that six-sidedness had something to do with packing them together efficiently. He was intrigued by the fact that hexagons appeared in other situations where efficient packing was paramount, as in the individual cells in a bee's honeycomb, or even the arrangement of seeds in a pomegranate. But he couldn't explain why snowflakes are six-sided.

It wasn't until well into the 1900s that at least part of Kepler's question could be answered. The architecture of snowflakes depends first on the water molecule itself. Water is, of course, two atoms of hydrogen and one of oxygen: H_2O. At room temperature, when water is liquid, the H_2O molecules are vibrating and sliding past each other, colliding and recoiling, leaving virtually no space between them. But if the temperature drops low enough, this normal jostling of water molecules relative to each other can be overcome by electrical forces acting among them, and they then all snap into fixed positions relative to each other. That's freezing.

When freezing immobilizes the water molecules, it forces them to move apart and take only rigid arm's-length positions with respect to each other. X-rays of such ice crystals reveal a remarkable repeating pattern of hexagons: six water molecules at each corner, in turn bonded to other water molecules beside, above and below. Kepler was closer than he realized: at the micro level, a mass of ice crystals looks like a vast honeycomb.

This repeating hexagonal pattern holds the key to snowflakes. Under the right circumstances in the atmosphere, a crystal will start to grow (usually by latching on to a microscopic dust particle), adding water molecules to its edges, always preserving the underlying hexagonal organization. By the time it's big enough to see, you have a snowflake.*

The greatest snowflake scientist of the twentieth century is the late Ukichiro Nakaya, a nuclear physicist whose life was changed when he had the (mis)fortune to be appointed to a chair of physics in Hokkaido, the north island of Japan, in 1932. The university there didn't have the facilities to do nuclear physics, but they did have lots of snow. Nakaya took the pragmatic approach and changed his field of study. He was the first scientist to produce snowflakes in the lab. He strung rabbit hairs like miniature clotheslines in cold chambers filled with water vapour. Rabbit hairs are lined with little nodules that provide the same opportunity for initiating the growth of snowflakes as do dust particles in the air: snow crystals can settle on them and begin to add new crystals from the surrounding air.

Nakaya showed for the first time why snowflakes can be anything from delicate, feathery six-fingered stars to blunt six-sided

---

* The rigidity and immobility that I've implied exists in an ice crystal is relatively true when you compare it to liquid water, but such a crystal is by no means totally inactive. There's evidence that the average water molecule in an ice crystal jumps out of its position once every *millionth* of a second, wanders a little way down the crystal, maybe a distance of eight molecules or so, then kicks another H_2O out of its place and jumps back into the crystal.

shields, or even the pillars with hexagonal caps on both ends that Descartes mentioned. Their shapes reflect their histories. As Nakaya wrote, "a snowflake is a letter to us from the sky."

Temperature and humidity are the key—as they change, the way snow crystals grow changes as well. If a snowflake is forming in fairly dry air at minus 15 degrees Celsius, it will be plate-shaped, but at 10 degrees less than that it will form as a solid column. The feathery six-armed Christmas-card snowflakes develop only in very wet air at around minus 14 degrees Celsius. You can imagine how complicated it can be if a snowflake starts at one temperature and humidity, then is carried by winds up, then down, first moving quickly, then slowly (differences in speed affect the growth of the crystal), then finally falling to earth, encountering yet more changes in temperature and humidity. With each change the growing flake will alter its ongoing pattern, while preserving what's already there. As Nakaya suggested, you could read the history of that flake by its shape.

Why should temperature and humidity make so much difference? If every ice crystal, at least at the molecular level, is exactly the same, why isn't every snowflake the same? This question is right at the frontiers of physics. Everyone agrees on the general picture: free-floating water molecules, or small groups of them frozen together, bump into the growing snowflake and lock into place. The higher the humidity, the more molecules competing for available spaces; the lower the temperature, the easier it is to form the solid bonds that hold them there. You can even imagine how a plate could change to a column—by growing up rather than out— and how a plate that grew only at its corners, rather than the edges,

would soon grow arms. But it's nearly impossible to know what
the exact conditions are in the microscopic neighbourhood of the
crystal, and that's where tiny differences in temperature or
numbers of water molecules might change one shape to another.
The flakes that Descartes saw obviously started out as columns
and suddenly changed to plates, yet these and almost all
snowflakes end up perfectly symmetrical—you don't find them
with one arm ten times longer than the others. But how does one
end of a snowflake know to grow at just the right speed that it will
balance what's happening on the other end? *That* may be the
toughest question of all.

Finally, is it really true that no two snowflakes are alike? Until
1988 the answer was yes; now it's yes . . . maybe. In May 1988,
Nancy Knight of the National Center for Atmospheric Research in
Boulder, Colorado, reported finding two snowflakes that were
"virtually identical." These flakes were collected on November 1,
1986, by researchers studying clouds in an airplane flying at about
6,000 metres (20,000 feet). They're shaped like tiny columns with
vase-shaped hollow centres and appear to be the same. Knight, an
experienced snow researcher, said it was the first time she had seen
flakes like this, "which, if not identical, are certainly very much
alike." Of course, it's impossible to know if these flakes were exactly
identical—there's only one photo from one angle at one magnifi-
cation. This leaves the door open for traditionalists to continue to
rely on some very convincing statistics to argue that even these two
probably weren't "identical."

Consider the number of snowflakes that have fallen on the
earth since the beginning of the planet. One estimate is 10^{35} flakes

of snow—that's 1 followed by 35 zeroes—a weight of snow fifty times the mass of the earth, even though each flake weighs only a millionth of a gram. Surely two of those would be the same? Well, you must take into account that each snowflake contains something like 10^{18} molecules of water, and those can be arranged in lots of different ways. What's lots? A rough calculation suggests that for an individual flake growing gradually as it falls, there might be a million occasions when water molecules have a choice of more than one place to attach. This makes the number of possible flakes incredibly huge, trillions of trillions of time greater than the total number of flakes that has ever fallen. Besides, how likely is it that two snowflakes would have exactly the same histories? Even two balloons released at exactly the same place at the same time travel different paths because of unpredictable breaths of air. And if that isn't enough, even snowflakes following exactly identical paths couldn't be the same, because the first one would absorb water molecules that then would be unavailable for the second.

The inescapable conclusion from the statistics is that it's over-whelmingly unlikely that any two snowflakes are identical. That being the case, where does Nancy Knight's photograph fit in? Either it's a picture of two flakes that are almost exactly the same, and so is a curiosity but not much more, or it really is the only existing record of two identical snowflakes. That would make that single photo a scientific masterpiece, the highlight of four hundred years of snowflake research.

Feathers

We've all, at one time or another, picked up a feather, maybe run a finger along the edge, tickled someone with it, then thrown it away. Just about everyone can describe a feather, too, but most never notice one of its most obvious features: the placement of the quill. It doesn't run down the middle of the feather. It's off to one side, the side that faced into the oncoming air when the feather was still attached to the bird. The effect is most pronounced in flight feathers, particularly those at the outer tips of the wings— the quill is almost right at the leading edge. Even the tiny barbs that stick out from the quill are stiffer along the leading edge. In wing feathers closer to the body this arrangement is less pronounced, and the feathers actually covering the body are nearly symmetrical.

This is not a fluke. There are very good aerodynamic reasons for having the quill, the main support of the feather, close to the leading edge. It's crucial that each individual feather is oriented properly at every moment in flight: on the downstroke the outermost flight feathers separate, each one angled downward, cutting into the air like a tiny wing, while on the upstroke these same feathers rotate so that air just slips through, offering no resistance as the wing is raised. The downstroke is the one that lifts the bird, and as the wing is brought down, that lift from the air is concentrated near the front edge of the wing—not, as you might have expected, in the middle. If the feather isn't reinforced near the leading edge, it'll twist or even buckle under the forces, and the bird will lose control. (The Germans found this out the hard way in World War I, when they produced a new single-wing aircraft with reinforcing spars located near the middle of the wings. Pilot after pilot found that pulling these planes out of a dive tore the wings off, and it was only when the designer, Anthony Fokker, moved the supporting spar to the front of the wing that the planes stayed airborne.)

You're in good company if you've never noticed the position of the quill. In one celebrated case, the experts missed it too. These experts were scientists engaged in a vigorous debate over the flying ability of an extinct bird called Archaeopteryx. The only traces of this animal in existence are six skeletons and a single fossil feather, but while the feather is a perfectly clear impression in rock, it took more than a hundred years for someone to say, "Let's look at the quill in that feather."

Archaeopteryx lived about 150 million years ago, right in the middle of the heyday of the dinosaurs. Its bones are exactly like those of a number of relatively common pigeon-sized hunting dinosaurs, animals that ran around on their hind legs chasing insects, catching them in their front claws and crunching them in toothy little jaws. But fanned out around the twisted bones of the fossil skeletons of Archaeopteryx are imprints of feathers—perfectly ordinary feathers. They're practically the only things that make these animals something more than just another dinosaur, and more than one scientist has suggested that if you took away the feathers, these bones would have been labelled "dinosaur," locked in a drawer and forgotten. In fact, this is exactly what happened to one of the most recent Archaeopteryx fossils, the discovery of which was announced in 1988. It had remained unrecognized in an amateur's fossil collection for years since its discovery; no one had noticed the faint feather traces around the bones.

But what kind of animal was a little two-legged carnivorous lizard that happened to have feathers too? Scientists think it's one of the first birds, an animal well on its way to birdom but still retaining its dinosaurian teeth and skeleton. (In fact, the Archaeopteryx skeletons provide evidence that the dinosaurs didn't die out as a failed evolutionary experiment—they continue to be successful as birds.) But if Archaeopteryx wasn't much more than a lizard with feathers, could it fly?

This is not a simple question, mostly because there are only nine highly squished skeletons around to provide the answer. What can be deduced from the twisted bones doesn't exactly make Archaeopteryx look like an aerial acrobat: it has only small areas on its bones for the attachment of flight muscles (so they couldn't have been very large or powerful), it has no breast bone to serve as the anchor for the big chest muscles that should provide most of the thrust in flight, none of the arm bones are fused together to provide the rigid structure needed to absorb the force of flapping, the pelvic girdle isn't strong enough to take the force of a typical landing, and one study has even suggested that if the rather weighty Archaeopteryx had dared to flap its wings, the wing bones would have snapped from the stress. The feathers don't even appear to be implanted securely in the wing bones as they are in modern birds, but instead are simply stuck into the skin. It would be, at best, a rickety flying machine.

But you have to account for the feathers somehow. Short feathers covering the body can be used as insulation—in fact feathers are just as good as down for that purpose—so if you buy the idea that the dinosaurs were hot-blooded, you can explain the body feathers that way. But what about the long feathers attached to the outstretched forelimbs? They aren't keeping anything warm, and scientists have ranged far and wide to explain them. John Ostrom of Yale University was one of the world's experts on Archaeopteryx, and he suggested in the 1970s that the long feathers on the wings helped the animal as it ran around looking for insect prey. He envisioned the wings as snares—Archaeopteryx would sprint around on its hind legs, holding out its feathered forelimbs, and either swat or envelop flying insects. Ostrom eventually abandoned this idea, but it at least opened the door to thinking about how feathers might serve purposes other than flight.

Some scientists now think Archaeopteryx might have hunted in shallow water the way the African black heron does today. It stands motionless with wings spread out, creating a little shady patch in the water. It's great for fishing. The fish flee into the shade, the shadow eliminates glare and so allows the heron to strike more accurately, and the dark background created by the wings may make movements of the heron's head harder for the fish to see. So feathers for shade is a possibility.

There are other explanations for Archaeopteryx's feathers, but for my money, one of the best observations in this whole debate is also one of the simplest. In 1978 two American scientists, Alan Feduccia and Harrison Tordoff, simply looked at the first Archaeopteryx fossil ever found: the single feather from 1860. The quill is obviously way off to one side, just like the wing feathers of birds today. Feduccia and Tordoff pointed out that feathers are like that only when they're used for flying, and so Archaeopteryx was probably a flying animal.

Modern birds that can't fly, like ostriches, have feathers with the quill running right down the middle. Other birds that are thought to have lost the ability to fly relatively recently in evolutionary time have feathers in which the quill is on its way to the middle but hasn't quite reached there yet. On the other hand, the better the flier, the more pronounced the tendency of the quill to lie near the leading edge of the wing. Falcons and hummingbirds practically have the quill coming right off the feather. So, Feduccia and Tordoff say, this must mean that Archaeopteryx, however inadequate its skeleton, must have been a flier—indeed, a pretty good flier. This is undoubtedly not the end of this story, but it is puzzling why it took until 1978 for someone to look at the 1860 feather and notice something that anyone could have seen with a glance. It's somehow reassuring that even the experts sometimes miss the obvious.

Unfortunately any clarity provided by the Feduccia/Tordoff discovery didn't last. Since 1978, many more species of fossil animals with feathers have been discovered. Many of them obviously couldn't fly: they are feathered ground-dwelling dinosaurs. Most of these have been found in the now-famous Liaoning fossil beds in China. In fact the Chinese fossils have changed our notion of feathers completely and utterly. It used to be that an animal with feathers was a bird. Now we know that many of them were just small carnivorous dinosaurs whose bodies were covered with feathers.

But the skeletons of these creatures show clearly that they couldn't fly—they have even fewer of the adaptations for flight that Archaeopteryx had. So what were their feathers for? The best guess is that most of them were short, flat-to-the-skin feathers that served as insulation. These dinosaurs were likely warm-blooded, and, like small-bodied animals then and now, needed to do everything they could to retain that warmth. It must have been only much later—millions of years later—that these feathers evolved to lend an aerodynamic advantage to those animals who invented flight.

Some of the fossil creatures from Liaoning, among them a selection of early fliers, went to extravagant lengths to clothe their bodies in feathers. Some had long flight feathers, not just on the forelimbs but on all four legs and along the length of the tail as well. Some even sported a crest of feathers on their heads, as blue jays do today. Unfortunately, it's a tricky business to infer the ability to fly simply from a sometimes incomplete skeleton and the rocky impressions of feathers. It's clear that feathers didn't evolve just for flight—they existed long before that. But exactly which was the first animal to fly is still an unanswered question.

But old Archaeopteryx hasn't been completely superseded by the amazing array of new fossils. In late 2005 a study of a new fossil of the animal showed that it shared even more anatomical characteristics with other small non-flying carnivorous dinosaurs, making Archaeopteryx, still considered the first bird, much more like those dinosaurs. Indistinguishable, even, except for its feathers. Feathers that clearly allowed this animal to fly.

THIS CHAPTER
IS A YAWNER

You'd think there would be only two important questions to ask about yawning: why do we do it, and why, when one person starts yawning, do others immediately follow? But scientists are much more cautious than that: they want to make sure they know what a yawn *is* first. And it is "a stereotyped and often repetitive motor act characterized by gaping of the mouth and accompanied by a long inspiration followed by a short expiration." That may be a long and accurate definition, but it makes yawning sound like breathing, and other than the fact that they both involve air going in and out, they couldn't feel more dissimilar. One gets the feeling that if that's the definition, then we're still barely out of the starting gates on this one.

How often do we yawn? Students sitting alone in a room for half an hour, pushing a button whenever they start to yawn and releasing it when they finish, have established the length of an average yawn to be six seconds, give or take two seconds. But amazingly, the frequency of yawning varies from only one in half an hour to *seventy-six*. If those seventy-six yawns each lasted six seconds, then that individual spent an entire half-hour yawning, with an average break of only seventeen seconds between yawns.

There's a popular idea that we yawn to get more oxygen to our brains, or conversely, to rid our blood of excess carbon dioxide. The more oxygen in the blood, the more glucose we can burn for energy. Carbon dioxide is a waste product that takes up space in our red blood cells that could be used for oxygen, and it could conceivably build up if we're breathing shallowly for long periods of time. This theory would explain why we yawn when we're tired or sitting in a stuffy, overheated lecture hall at three o'clock in the afternoon—not enough fresh air.

But one of the few psychologists researching yawns, Robert Provine at the University of Maryland, has found that changing the levels of carbon dioxide or oxygen seems not to affect yawning. Breathing *pure* oxygen didn't make subjects yawn less, and breathing air high in carbon dioxide didn't make them yawn more. Now it may be that setting this up in a laboratory situation somehow misses a critical factor, but if the cause of yawning were simply the levels of gases in the blood, experiments like this should show some effect. And they don't.

But these experiments don't completely kill the idea that yawning might refresh an oxygen-deprived brain. It's possible that by forcing your mouth wide open as you yawn, and stretching at the same time, you constrict some blood vessels while dilating others, with the net effect of forcing more blood to the brain. More blood means more oxygen, and presumably heightened alertness. It is true that opening your mouth wide is an important part of yawning: volunteers who yawned with clenched teeth reported it to be a highly abnormal experience that did nothing to satisfy the urge to yawn. If you try this yourself you'll see how incredibly unfulfilling it is, and the difficulty of explaining why that should be so just underlines how little yawning is understood. After all, you can still breathe in and out when your teeth are clenched, so it can't be lack of fresh air that leaves you dissatisfied. Maybe it's true that the wide-open stretching of your jaws and the resulting contraction of the facial muscles really do cause changes in blood flow to the brain. But does that mean that instead of yawning, you could just open your mouth as wide as possible for a few seconds, breathe in and out and get the same effects and the same satisfaction? No, it wouldn't be the same.

Medical doctors have collected evidence that suggests that yawning is controlled in a very primitive part of the brain, an area that runs for the most part on automatic pilot, beyond the reach of our conscious mind. Anencephaly is a tragic birth defect in which a newborn is missing most of the brain. Such children (who don't usually live long) have no cerebral hemispheres, the convoluted upper parts of the brain where most of our thinking takes place. But they have no trouble yawning. This must mean that the control of yawning is in one of the few brain parts they have—a lower, more "primitive" area like the medulla oblongata, a structure at the bottom of the brain that's responsible for normal breathing. It seems reasonable that yawning would be pretty well out of reach of your conscious mind; not only can a yawn start without a thought from you, it's also so difficult to stifle one once it's started—a powerful yawn can easily override your conscious efforts to stop it.*

This yawning centre must have a complicated set of connections within the brain, because it has to trigger yawns in response to all kinds of different stimuli, ranging from the sight of other people yawning, through fatigue to boredom. In fact the absence of yawning may indicate a problem not with the yawning centre itself but with another part of the brain to which it is connected. Psychiatrist Hans Lehmann at McGill University noted that he almost never saw mental patients yawn—on the wards, on buses, at public gatherings or in restaurants. It made perfect sense to Lehmann that these patients, most of whom were schizophrenics, didn't yawn because they were so emotionally flat that they had little interest in what was going on around them. He reasoned that to be bored enough by what's going on around you to yawn, you have to be aware there *is* something going on. These patients were too withdrawn to know what was going on around them, so they couldn't be bored, and didn't yawn. It is said to be a positive sign when some mental patients yawn, because it suggests they're trying to establish contact with reality.

So why do we do it? Ronald Baenninger of Temple University has pulled together evidence that suggests to him that we yawn to arouse the nervous system in anticipation of an increase in activity.

---

* There have also been cases where a patient with an arm paralyzed by a stroke or other brain damage has miraculously moved that paralyzed limb while yawning! The patient has no control over the movement: the arm just stretches on its own as it would normally do to accompany yawning. This strange occurrence suggests a link between yawning and stretching that completely bypasses the normal voluntary control of movement.

He had volunteers wear wrist bracelets that record the movement of the wrist (and serve as a proxy for overall activity) and at the same time keep a diary of their yawning. The times when these people yawned the most coincided with significant activity changes, such as when they went to bed or got up in the morning. Baenninger's theory is perhaps unnervingly flexible in that it can go in both directions to explain why people yawn when they wake up—they're about to ramp up to start the day—but also when they're readying for bed—in this case they are trying to stave off the impending lack of activity. But overall there seems to be a consistent sequence: first comes the yawn, then the increase in activity.

Baenninger found that the subjects in one of his studies yawned less during the week than on weekends, even though they were getting less sleep during the week. Not only do those data cast doubt on the idea that we yawn because we're sleep-deprived, they also support the notion that the need for arousal may fluctuate from day to day. In this case, students were the experimental subjects, and they actually did yawn more during the week, especially when attending long lectures on weekday afternoons. They had no need for that kind of arousal on the weekend. Other studies have shown that driving, characterized by periods of intense arousal followed by periods of equally intense boredom, encourages very frequent yawns.

The biggest mystery of all about yawning is its infectiousness. You see somebody yawn, you yawn yourself. This is sometimes explained as coincidence—everyone is yawning at the same time because there's too little air circulation or too little oxygen in the air, and therefore it only appears to be contagious. But as I've already pointed out, experiments have shown that breathing air that's too high in carbon dioxide has no effect on the frequency of yawning. More to the point, you don't even have to be in the same room as the yawner to feel the yawning urge. Students in Robert Provine's lab who watched videotapes of an actor yawned much more often if the actor was yawning than if he was smiling (55 percent versus 24 percent).

Actually you will be prompted to yawn if you just read about yawning. (As you've probably already noticed.) The sound of yawning even provokes blind people to do it. We proved the power of this contagiousness on our program "Quirks and Quarks" in December 1988, when after making a few brief comments on some yawning research, I suggested that any listener who had felt the urge to yawn while I had been talking should write to us. I expected at most a couple of dozen letters (we weren't giving away T-shirts or promising to mention anyone on air), but we were all stunned by the avalanche of letters. The final count was close to three hundred, and the picture that emerged was that of thousands of Canadians yawning uncontrollably at about 12:50 P.M. on December 17, 1988, as they listened to the radio.

There have been some recent attempts to figure out exactly what's going on in our brains when we yawn in response to someone else. One of these put people in an MRI and watched to see which brain areas became activated as they saw a yawning face. The researchers, a team of Finnish and German scientists, were curious to see if the area of the brain containing "mirror" neurons was activated. These neurons are involved when we deliberately mimic the actions of others, and are thought to be absolutely crucial for learning how to improve your golf swing, learn sign language or any one of a hundred other movement-based skills. However, because the urge to yawn is so automatically contagious, the scientists suspected that it might bypass this network that is responsible for conscious, deliberate imitation. And they were right: the brains of people who had just witnessed yawning faces experienced a sudden increase in activity on both sides of the brain in a place called the superior temporal sulcus, a patch of neural tissue just around the ear. (There was no such activation when the faces were simply moving their tongues with their mouths open.) The superior temporal sulcus is known to be intimately involved in responding to social cues. So far, so good. In addition, the area of the brain that contains the mirror neurons, just under the temple on the left side, did not respond to yawning faces. So it appeared from this study that the contagiousness of yawning is an automatic—and likely ancient—response.

Funnily enough (and this should be a caution to anyone who puts too much stock in any single study) a separate piece of research identified different brain structures that were activated by the sight of yawns, and these were areas known to be involved in what's called "self-referential" thinking—"all about me" thinking. This includes remembering things that have happened to you, identifying words that refer to you and monitoring your own actions and words. Where these two studies share common ground is that they both point to the contagion of yawning as being a primitive, automatic response—a form of unthinking empathy.

But why would we respond to the yawns of others with one of our own? The most popular suggestion—and one that's backed up by those brain-imaging experiments—is that yawning might be a hangover from our ancestral animal past, something called a "stereotyped action pattern." This is an automatic response to a signal that is common among birds and animals— for example, a frog will lunge hungrily, and unthinkingly, at any small dark moving object—but why would we yawn simply because other humans are yawning? In the early sixties, anthropologist Ashley Montagu, who was one of the believers in the idea that yawning heightened alertness by shunting fresh blood to the brain, speculated that group yawning would, by this mechanism, increase everyone's alertness. On the other hand, in a beautiful example of how any theory fits if you have very little data, an English doctor named Malcolm Weller had recently suggested the exact reverse—that the infectiousness of yawning might have originated as a signal among social animals to go to sleep, not to revive. Weller thinks that if one animal were to signal his weariness by yawning, the pack would imitate him and soon settle for the night.

This picture of our distant ancestors engaging in communal yawning for the good of the group is fine, and if that behaviour became automatic, we might well still be doing it today, long after it had outlived its usefulness. But this is a tricky business, using what anthropologists *think* our ancestors were doing to explain why we do it today.

If our primitive ancestors used yawning as a signal, at what point did we start covering our mouths when we yawn?

Obviously it would have been counterproductive for those animals that were signalling each other with their yawns to have been hiding those same yawns with their hands or paws. You can argue that covering the mouth is a recent human cultural invention superimposed on the ancient habit of yawning openly. But by doing so, you're ignoring some intriguing observations of our closest living relatives.

By studying the habits of modern apes and monkeys, scientists believe they can infer some of the ways our ancient ape-like ancestors behaved millions of years ago. Yawning is a social signal among animals like monkeys and baboons, but an aggressive one. Dominant male baboons engage in "threat yawns," a gesture some scientists interpret as being more threatening than a raised eyebrow or a stare. Anthropologist John Hadidian spent seven hundred hours watching black apes yawn, and he concluded that the ones doing the yawning are the dominant males: the number-one male averaged three yawns an hour, while number four yawned only once every three hours. These apes, and many other related species, have especially long canine teeth, which a yawn displays to great advantage.

Some biologists have even reported seeing subordinate animals cover their yawning mouths with their paws, apparently to prevent those yawns from being interpreted as threats they couldn't back up. We claim that we cover our mouths because to do otherwise would be "rude," but maybe that's just our civilized way of saying that it's somehow threatening. An open-mouthed yawn directed at you may not actually be a physical threat, but it can convey an unflattering, even unfriendly message (even without prominent canines).* What is really needed here is a study of human groups to see if the dominant individuals don't bother covering their mouths, while the submissive ones do. The Godfather's henchmen always laughed when he did, and stopped laughing when he did. What did they do when he yawned?

And when all the research falls into place, and we know exactly why we yawn, and especially why we yawn when others are yawning, where will the rat data fit in? The rat data? A variety of studies have established that certain drugs will quickly bring on two responses in rats: yawning and erections.

---

* Just to add another twist to the story, Robert Provine's research has shown that people will yawn in response to a yawning face with the mouth blanked out—the eyes and forehead are enough to stimulate yawns. So covering your mouth when you yawn does not interfere with its contagiousness.

Just Walkin'
in the Rain

Does it seem obvious to you that if you're caught in a heavy rain, you should run as fast as you can to your destination to minimize the amount of rain that falls on you? The less time, the less rain? But it's not that simple. By going faster you're actually running into drops that otherwise would have fallen harmlessly to the ground in front of you. Aha, you say, but you're also leaving behind drops that would have hit you. True, but what if the rain is slanting, either in your face or in your back, what then?

Fortunately a handful of physicists and mathematicians have taken this problem seriously. Their analyses produce different numbers, but they agree on some strategies for staying as dry as possible if you're forced to walk in the rain.

If you're going about this scientifically, you never do your calculations using a real human body—you "model" the problem by using a geometric shape that approximates a body and makes the mathematics a lot easier. Alessandro De Angelis of the Institute of Physics at the University of Udine in Italy created what he calls a "rectangular parallelepiped" (which looks a little like a cereal box) to serve for his human; S. A. Stern of the University of Texas was content with a "square planar surface element" for his. Both these geometric representations of humans are programmed to move horizontally through some imaginary space, with rain pounding down on them from above—a physicist's rain, of course, with uniform drops falling in a uniform way ("a horizontally isotropic and time-invariant rainstorm of uniform intensity N drops per second per unit area," as Stern puts it). Of course, both physicists make the point that the falling rain isn't the only concern: there are also the raindrops that you collide with when you're running. (Just as, when you first step into the shower, the initial drops that wet you are drops you walk into, not drops that fall on top of you.) Those drops constitute a second, important source of wetness to work into the mathematics.

Stern's analysis shows that the speed with which you move will not affect the number of drops hitting you from above but can influence the number that you encounter coming at you from the front, the drops that you run into as you move forward. The faster you go, the fewer of those drops you'll meet. You may feel that more rain is striking you, because of an increase in the number of drops per unit time, but the total number of drops will be lower if you hurry. Stern points out that the best thing to do is minimize the surface area you present to the oncoming rain by leaning as far forward as you can. The ideal, he says, would be to lie down and ride a skateboard through the rain, but if that's not practical, you could shuffle as fast as possible sideways, leaning as far as you can in the direction you're going.

De Angelis, on the other hand, arrives at the same conclusions, but with a slightly different, and very continental, attitude.

His paper in the *European Journal of Physics* is entitled "Is It Really Worth Running in the Rain?", and while he agrees with Stern that the faster you go, the less wet you get, the numbers he comes up with suggest that walking at a brisk walk (three metres per second) will get you only 10 percent wetter than if you were running at sprinter-like speed (ten metres per second). As he says, "the benefit does not justify the supplementary effort." One gets the impression De Angelis would prefer to duck in for an espresso and wait for the rain to stop.

What if we go beyond the tightly constrained laboratory model, the physicist's ideal rainstorm? Further calculations have been made showing that as raindrops get larger, fewer of them hit you, but their size ensures that they dump more water on you. And if a wind is blowing, you've got to take the analysis a little further. It's still technically, if not practically, correct to move fast if the rain is blowing into your face or coming from either side, but if the rain is blowing into you from behind, you have to make an adjustment. Running at high speed in this case will only get you wetter, because you'll be catching up to and smashing into raindrops. You have to try to match your speed to the speed that the falling drops are drifting into you from behind. The greater their horizontal speed, the faster you're going to have to run.

If you fail to adjust your speed carefully to these drops, you may get up to four times wetter than you need to. But as Dr. De Angelis pointed out, there's a limit to how fast you should expect someone to run to stay somewhat drier. Mathematician David Bell calculated that to keep pace (and so stay drier) with a tailwind that's slanting at 45 degrees into your back, you'd have to run at a four-minute-mile pace. Only you can decide whether it's worth it.

There is, of course, one rainstorm where you could theoretically stay perfectly dry. If the rain is being blown so hard that it's moving parallel to the ground, and you could keep up with it, then you run along with the rain and never get wet. You'd never catch up to the drops ahead of you, the drops behind couldn't catch you, and nothing would fall on your head because they're all travelling horizontally. But the moment you stopped, you would be soaked.

In most circumstances, then, the bottom line is to run as fast as you can (or want to) in a storm and you'll stay drier, except in the special case when the rain is slanting into you from behind. Then you must match your speed to the horizontal speed of the rain. This is a perfect example of the beauty of the science of everyday life. You begin with line after line of calculations in esoteric journals, but you end up with drier clothes in a storm. Of course, if your body deviates here and there from a "square planar surface element" or a "rectangular parallelepiped," you can't blame the mathematicians if you get wetter than the theory predicts.

Unsteady As
She Goes

For all that baseball fans like to brag about the subtlety of their game, they're still seduced by the obvious: upper-deck home runs and ninety-five-mile-per-hour fastballs. At the same time, a pitch like the knuckleball is generally dismissed as a freakish delivery that is the last resort of a middle-aged, pot-bellied pitcher who has lost his fastball. Yet the knuckleball is a unique pitch, one that defies understanding by hitters and scientists alike.

The standard baseball pitches—slider, curve and fastball— slavishly follow the laws of nature that dictate how a ball can curve. The knuckleball doesn't. Normally these laws make it possible to predict where a ball will go and how it will get there. You can't do that with the knuckleball.

It is at first sight an unspectacular pitch. A knuckleball pitcher doesn't fire the ball to the plate, he lobs it. But while you'd expect that would make it vulnerable, the knuckleball compensates for its lack of speed by moving in strange ways. It can dip suddenly, or dart to one side or the other, or when you least expect it, float straight in over the plate. It is such an unpredictable pitch that Paul Richards, who was the regular catcher for the great knuckler Hoyt Wilhelm, invented a huge catcher's mitt just to ensure he could get a little piece of the ball as it went by. Even though Richards *knew* a knuckler was coming, he didn't have a clue where it was going to go. To understand why the knuckleball is so perverse, you have to begin with something straightforward: the reasons why a baseball curves if it is spinning.

The first person to experiment with the curving flight of spinning objects was Benjamin Robins, an English scientist (or "natural philosopher" as they were called then) of the early eighteenth century. He published a book in 1752 called *New Principles of Gunnery*, and in it described some ingenious experiments that showed that musket balls deviate from a straight line, depending on their spin—exactly the phenomenon that baseball pitchers exploit with every pitch.

Robins's experiments were demonstrations on a grand scale. He set up a musket range in an English garden, with a stone wall at one end as the target. He then anchored a musket in a fixed position three hundred feet from the wall and aimed at it. Between the musket and the wall, Robins erected two giant screens of extremely thin tissue paper, one just fifty feet from the musket, the other fifty feet farther. Both screens were directly in the line of fire. He then fired a series of one-ounce balls from the musket, with a crowd of curious spectators looking on.*

Each shot punctured first one screen, then the next and finally smashed into the wall. By measuring how far the punctures in the

---

* The context of Robins's report suggests that these onlookers were probably potential skeptics, for he expressed the hope that having performed the experiments in this "publick manner . . . will secure me from the harsh and malevolent censures which propounders of new opinions are generally exposed to."

screens (and the mark in the wall) lay to the right or left of a straight line, Robins could record the flight path of each ball. Having done the experiment many times without an audience (he was no fool), he knew what to expect. The final impact point of the musket balls on the wall at the end of the garden was as much as 39 inches to one side or the other of the initial puncture mark in the first screen. That's a curve of 39 inches over a distance of 250 feet, which, curiously enough, is roughly equivalent to a 10-inch curvature over the distance from the pitcher's mound to the plate, well within the range of curves modern researchers have measured for baseballs.

Not only was Robins pretty sure of what was going to happen before he brought the crowd in to watch, he also had a theory to explain it. He was convinced a musket ball had to be spinning before it would curve like that, and he suspected that the ball acquired spin as it rubbed against that the barrel on its way out. He also figured that a clockwise spin (as viewed from above) made a ball curve to the right, counter-clockwise to the left.

So, with an unerring sense of showmanship, Robins bent the barrel of a musket 3 or 4 degrees to the left, aimed it at the first screen and fired. You'd expect that a musket pointing left would send the ball to the left, but Robins boldly predicted to the crowd of onlookers that the ball would "incurvate toward the right." He was counting on the fact that the ball would roll along the right-hand side of the bent barrel, acquiring so much clockwise spin that it would, by the time it reached the wall, hit to the *right* of centre. And it did so in dramatic fashion. Even though the ball from this crooked barrel hit the first screen an inch and a half to the left of a previous shot, and was still three inches left at the second screen, it had curved—as predicted—fourteen inches to the right by the time it hit the stone wall. Robins does not record the reaction of his peers, but they must have been impressed.

Not only does Benjamin Robins deserve recognition as the first person to actually perform experiments demonstrating the curving flight of spinning balls, he also had a pretty good idea of why these balls curved as they did. He suggested that a spinning ball creates a kind of whirlwind around it—on one side of the ball the whirlwind and the onrushing air are moving in the same direction, but on the other side they're opposed. The ball curves toward the side where the whirlwind and the outside air are moving in the same direction.

Today, English gardens have given way to the wind tunnel, and physicists can suspend and spin baseballs (and other objects) in streams of air moving at any speed they like. Everything that happens can be photographed, videotaped and analyzed. It turns out that Robins's idea of a whirlpool around a spinning musket ball wasn't far off—there *is* a very thin layer of air that's stuck to the baseball and whirls with it as it spins. It's called the boundary layer, and it clings to the surface of the ball by friction. The rougher the surface of the ball, the thicker the boundary layer. (I'm not sure what happens to the boundary layer when the ball is sitting in the umpire's pouch before being put into play—no one's done any experiments to find out.)

The boundary layer is the secret to all baseball pitches. Imagine you are hovering over a baseball spinning clockwise as it moves from the pitcher's mound toward the plate. On the left side, the ball and its boundary layer are turning into the onrushing air, but over on the right, the outside air and the boundary layer are moving together. The air on that side will slip past the ball smoothly. But on the left, where the boundary layer is turning into the oncoming air, there's trouble. The two airs clash, creating turbulence—the air gets chopped up—and as a result the boundary layer, instead of hugging the surface of the ball, is ripped away. When the boundary layer is torn away from one side of the ball, it leaves behind a rough wake like that which follows an outboard motor. The ball is then pushed in the direction away from the wake, in this case toward the right side, where there's no such turbulence.

So if a ball is spinning clockwise as you look down on it, it'll curve from left to right. If it's got overspin, it'll curve down: that's the major-league curve ball. If it has underspin, it'll stay up longer, which might account for the so-called rising fastball. The boundary layer of air around a baseball accounts for all the standard pitches, and the behaviour of each pitch is predictable, just as the flight paths of Benjamin Robins's musket balls were. But then what about the knuckleball—how can you explain a ball that is totally erratic and unpredictable? What loophole does the knuckleball exploit?

It's really very simple: musket balls, fastballs and curves all spin. The knuckleball hardly spins at all. It is actually slightly misnamed; the ball is held with the fingertips, not the knuckles. And unlike most pitches, it's thrown with a stiff wrist to minimize the amount of spin—some pitchers liken throwing the knuckle-ball to squeezing a watermelon seed out from between your fingers. The lack of spin causes the baseball to do strange things. Dr. Robert Watts at Tulane University put baseballs in a wind tunnel and spun them in various ways, with winds rushing past them, trying to determine what was unique about the knuckler. He found, to his surprise, that if the ball was rotated very slowly in the tunnel, it experienced different sideways forces. A push to the left that gradually intensified as the ball was slowly rotated would suddenly switch and push the ball to the right. Watts concluded that the stitching on the baseball was the secret.

There are 216 stitches on the ball, arranged in a continuous figure-eight, and as topographic features on an otherwise fairly smooth cowhide, they're the equivalent of mountain ranges. However, they are irrelevant if the baseball is spinning rapidly, because the stitches seem a blur to the oncoming air, and the ball behaves as if its surface is relatively smooth. But a baseball rotating very slowly on its way to the plate presents a different arrangement of stitches to the oncoming air at each moment. In fact, from the aerodynamic point of view, the knuckleball is virtually a different *object* from one second to the next, and there's no telling when the airflow around the ball will suddenly change, pushing it abruptly one way or another.

Robert Watts calculated that the ideal knuckler should make only one-quarter turn on its way from the mound to the plate. That should maximize the chance that some aerodynamic surprise will occur. Watts has seen those sudden and unpredictable forces develop in the wind tunnel when the ball is in the right positions, and has even witnessed a peculiar flip-flop force that he concluded must occur when the boundary layer of air is separating from the ball exactly where the stitches are. When this happens, the airflow flips back and forth from one side of the stitches to the other, giving the ball a little kick with each flip.

So Robert Watts attributes the knuckleball's unpredictability to the asymmetry of the stitches and the imbalances in airflow they create. But the knuckleball may not have yielded all its secrets yet. Other scientists think that the knuckleball changes direction abruptly because it is travelling at just the right speed to experience what is called the "drag crisis." That occurs when the air flowing over the ball changes abruptly from a smooth "laminar" flow to a turbulent one, or vice versa. Some researchers have even suggested that the knuckleball hovers so close to the drag crisis that it might even have air flowing smoothly over one part of the ball but turbulently over another. If that happens, the ball would undoubtedly be tossed abruptly to one side or the other.

In the end, does it really matter whether the knuckleball moves erratically because of the stitching or because of the drag crisis? Maybe not for a player—you can't hit the pitch any better knowing *why* it's unpredictable. But it should matter to the true fan, because the knuckleball surely is the most beautifully bizarre of any base-ball phenomenon. Where other pitches merely obey the straight-forward forces exerted on a spinning sphere, the knuckleball shuns mainstream aerodynamics and exploits fringe areas where strange and exotic behaviours can be found. Hard sliders and wicked curves are pitches that Benjamin Robins would have understood and appreciated. But even he, like thousands of frustrated batters, might have had a lot of trouble understanding the knuckleball.

Further Reading

I ronically, the problem with the science of everyday life is that the articles written about these phenomena appear in scientific journals that are anything but everyday reading. The following, however, are a few of the available books that deal with everyday science:

Brancazio, Peter J. *Sport Science*. Simon and Schuster, 1984. This book is strictly about the physics of sports, and while sports buffs may find there's too much physics, it's the only book of its kind.

Schrier, Eric, and William Allman, eds. *Newton at the Bat*. Charles Scribner's Sons, 1984. This book is about the science of sports, not just the physics, but it covers the subject in much less depth than Brancazio's book.

Walker, Jearl. *The Flying Circus of Physics with Answers*. John Wiley and Sons, 1977. Jearl Walker may have earned a reputation as something of a clown prince of physics as a result of his stage show and skits on "Quirks and Quarks," but this book contains more information, all of it solidly referenced, than any comparable book on the market.

Welcome to the Tongue-Show

A fascinating area of research with very little reading material available.

Ekman, Paul. *Telling Lies: Clues to Deceit in the Marketplace, Politics and Marriage.* W. W. Norton, 1985. This has nothing to do with tongue-showing, but if you're interested in the subtlety of facial communication, it's the book to read.

Smith, John, Julia Chase and Anna Katz Leiblich. "Tongue-Showing: A Facial Display of Humans and Other Primate Species." *Semiotica*, vol. 11, no. 3 (1974), pages 201–46. This is the best article on the subject, if you can find it.

The Roadrunner Triumphs Again

Kozhevnikov, Maria, and Mary Hegarty. "Impetus Beliefs as Default Heuristics: Dissociation Between Explicit and Implicit Knowledge about Motion." *Psychonomic Bulletin and Review*, vol. 8, no. 3 (2001), pages 439–53. Even physicists use impetus theory.

McCloskey, Michael. "Intuitive Physics." *Scientific American*, vol. 248, no. 4 (April 1983), pages 122–30. McCloskey reveals his experiments showing that many of us are medieval when it comes to understanding how things move.

Peary, Danny, and Gerald Peary, eds. *The American Animated Cartoon: A Critical Anthology.* Clarke, Irwin, 1980. The chapter titled "Meep-Meep" in this book gives you all the background you need on the Roadrunner and the Coyote. But to get a real feeling for the physics of the cartoons, you're best to watch the video *The Bugs Bunny Roadrunner Movie.*

"When the Moon Hits Your Eye . . ."

There is still so much controversy about the moon illusion that an entire book on the subject—full of competing theories—exists. It is:

Ross, Helen, and Cornelis Plug. *The Mystery of the Moon Illusion.* Oxford: Oxford University Press, 2002.

Kaufman, Lloyd, and Irvin Rock. "The Moon Illusion." *Scientific American*, vol. 207, no. 1 (July 1962), pages 120–30. There is no single article that presents all points of view, but this one is good on the apparent-distance theory.

Sights and Sounds in a Cup of Coffee

Schaefer, Vincent. "Observations of an Early Morning Cup of Coffee." *American Scientist*, vol. 59 (September–October 1971), pages 534–35. The original paper reporting the appearance of convection cells on the surface of a hot cup of coffee.

Senders, John. "The Coffee Cup Illusion." *American Journal of Psychology*, vol. 79 (1966), pages 143–45.

Walker, Jearl. "The Amateur Scientist." *Scientific American*, vol. 237, no. 5 (November 1977), pages 152–60. Walker investigates the physics of both tea and coffee, but I would be wary of his conclusions regarding the cooling rate of coffee with or without the cream. Those data have since been revised.

I made no reference to the coffee cup illusion in this chapter, but it's worth trying. If you stare into a black cup of coffee that's illuminated only by a single light directly above it, you'll see an image of that light reflected from the surface of the coffee. Position your head so that the image fills the cup, then suddenly move your head closer. The light appears to shrink and recede dramatically, rather than getting bigger as you might expect. Apparently the image of the light in your eye actually does get slightly bigger as you move your head closer, but the image of the cup increases proportionally much more. Your brain knows that the cup hasn't literally become bigger, so it explains the situation by assuming that the light has shrunk.

Sex and the Single Armrest

Fast, Julius. *Body Language*. Pocket Books, 1970. Scientists may shudder at the sight of "pop psychology" books like this, but chapters 3 and 4 have some interesting anecdotes about personal space.

Hai, Dorothy, Zahid Khairullah and Nancy Coulmas. "Sex and the Single Armrest: Use of Personal Space during Air Travel." *Psychological Reports*, vol. 51 (1982), pages 743–49. This is the paper describing the battle over the armrest in the economy-class section of airplanes.

Sommer, Robert. *Personal Space*. Prentice-Hall, 1969. This book ranges far and wide over the topics of personal space and territory, discussing not only personal space and its invasion but also the peculiarities of spacing in specific settings, including taverns in Edmonton in the mid-sixties. Anyone who was there at the time, as I was, can't help but agree with Sommer when he notes (on page 129): "the oft-described pattern of 'nursing' a single beer for an indefinite period is extremely rare in the Edmonton beer parlor."

Two Good Reasons for Having a Bath

McDonald, James. "The Coriolis Effect." *Scientific American*, vol. 186, no. 5 (May 1952), pages 72–78. This isn't a bad description of the Coriolis effect, written as it was when *Scientific American* was still understandable.

Rorres, Chris. "Archimedes." www.math.nyu.edu/~crorres/ Archimedes/contents.html. Mathematician Chris Rorres is the best source on all things Archimedes.

Shapiro, Ascher. "Bath-Tub Vortex." *Nature*, vol. 196, no. 4859 (December 1962), pages 1080–81. The Shapiro paper is the original experiment, the Trefethen paper that follows is the Australian replication.

Trefethen, L. M., et al. "The Bath-tub Vortex in the Southern Hemisphere." *Nature*, vol. 207, no. 5001 (September 1965), pages 1084–85.

No Redeeming Features

Barnard, Chris. "Birds of a Feather." *New Scientist*, vol. 83, no. 1172 (September 13, 1979), pages 818–20. This is the only popular article that includes some of the research on the behaviour of house sparrows in flocks. The rest is unfortunately still confined to esoteric science journals.

Kastner, Joseph. "The Great Sparrow War." *Smithsonian*, vol. 17, no. 8 (November 1986). This article details the public battle over the introduction of the house sparrow in the United States.

Taverner, P. A. *Birds of Canada*. Musson, 1945, pages 366–68. This book can be found only in the library or at second-hand bookstores, but it's worth reading the section on the house sparrow to experience Taverner's undisguised hatred of the poor bird.

The Dynamics of the Cocktail Party

Haykin, Simon, and Zhe Chen. "The Cocktail Party Problem." *Neural Computation*, vol. 17 (2005), pages 1875–1902.

MacLean, William R. "On the Acoustics of Cocktail Parties." *Journal of the Acoustical Society of America*, vol. 31, no. 1 (January 1959), pages 79–80.

Déjà Vu All Over Again

Brown, Alan S. *The Déjà Vu Experience*. Psychological Press, 2004.

Brown, Alan S. "A Review of the Déjà Vu Experience." *Psychological Bulletin*, vol. 129, no. 3 (2003), pages 394–413.

Sweet Thoughts from a Tiny Brain

Giurfa, Martin. "Cognitive Neuroethology: Dissecting Non-elemental Learning in a Honeybee Brain." *Current Opinion in Neurobiology*, vol. 13 (2003), pages 726–35.

Gould, James. "Do Honeybees Know What They're Doing?" *Natural History*, vol. 88, no. 6 (June–July 1979), pages 66–75.

Although this article predates some of the work suggesting that bees have mental maps, it still covers a lot of ground.

Maeterlinck, Maurice. *The Life of the Bee.* George Allen & Unwin, 1922. Maeterlinck saw in the bees a miniaturized version of human society, and while his language is a little quaint, he asks many of the same questions about bee intelligence that scientists are still asking today.

von Frisch, Karl. *The Dance Language and Orientation of Bees.* Harvard University Press, 1967. Much of the original research on bee dancing was published in German. Von Frisch relates those experiments in English in this book, reason enough for having a look at it. But it is worth reading anyway—it's the classic book on bee behaviour.

Winston, Mark. *The Biology of the Honey Bee.* Harvard University Press, 1987. This book may be more than you'd ever want to know about bees, but it covers all aspects of bee life.

The Big Bang on Cable

There are too many popular-astronomy books with material on the cosmic background radiation to list, but one (though a little dated and a little dense) is still the best:

Weinberg, Steven. "The Cosmic Microwave Background Radiation." Chap. 3 in *The First Three Minutes*. Basic Books, 1977.

I Think, Therefore I Blink

Bristow, D., et al. "Blinking Suppresses the Neural Response to Unchanging Retinal Stimulation." *Current Biology*, vol. 15 (July 26, 2005), pages 1296–1300. The laser light/brain stuff.

Ponder, Eric, and W. P. Kennedy. "On the Act of Blinking." *The Quarterly Journal of Experimental Physiology*, vol. 18

(1927), pages 89–110. This is the article where male and female blinking rates were compared on streetcars and in libraries. It's difficult to find, but reasonably easy reading. Immediately following this article is another on the blinking rates among different animals.

Stern, John. "What's behind Blinking." *The Sciences*, vol. 28 (November–December 1988), pages 42–44. At the time of writing, this was the only popular account of current research into blinking.

Stern, J. A., L. C. Walrath and R. Goldstein. "The Endogenous Eyeblink." *Psychophysiology*, vol. 21, no. 1 (January 1984), pages 22–33. This is the best review of modern blinking research, but beware: it's thick with jargon.

Dog Daze

Payne, C., and K. Jaffe. "Self Seeks Like: Many Humans Choose Their Dog Pets Following Rules Used for Assortative Mating." *Journal of Ethology*, vol. 23 (2005), pages 15–18. One of the two reports showing that dogs resemble their owners.

Roy, M. M., and N. J. S. Christenfeld. "Do Dogs Resemble Their Owners?" *Psychological Science*, vol. 15 (2004), pages 361–63. The other.

Sheldrake, Rupert. "Commentary on a Paper by Wiseman, Smith and Milton on the 'Psychic Pet' Phenomenon." *Journal of the Society for Psychical Research*, vol. 63 (October 1999), pages 306–11.

Wiseman, Richard, et al. "Can Animals Detect When Their Owners Are Returning Home? An Experimental Test of the 'Psychic Pet' Phenomenon." *British Journal of Psychology*, vol. 89 (1998), pages 453–62.

A Whiff of Asparagus

All the medical/scientific articles relating to this phenomenon are quite technical, but the two most recent are:

Lison, M., et al. "A Polymorphism of the Ability to Smell Urinary Metabolites of Asparagus." *British Medical Journal*, vol. 281 (December 20–27, 1980), pages 1676–78.

Mitchell, S. C. "Food Idiosyncrasies: Beetroot and Asparagus." *Drug Metabolism and Disposition*, vol. 29, no. 4 (2001), pages 539–43.

Mitchell, S. C., et al. "Odorous Urine Following Asparagus Ingestion in Man." *Experientia*, vol. 43 (1987), pages 382–83.

As a reminder of what a great character Babe Ruth was, try:

Creamer, Robert W. *Babe: The Legend Comes to Life.* Simon and Schuster, 1974.

The Science of Walking

Lovejoy, Owen. "Evolution of Human Walking." *Scientific American*, vol. 259, no. 5 (November 1988), pages 118–25. This article is a good indication of how far the scientific thinking on the origins of walking advanced in the twenty years since Napier's article.

Napier, John. "The Antiquity of Human Walking." *Scientific American*, vol. 216, no. 4 (April 1967), pages 56–66. The earliest evidence of upright walking is now two million years earlier than that cited here, but this is still a good review of the mechanics of walking.

Srinivasan, Manoj, and Andy Ruina. "Computer Optimization of a Minimal Biped Model Discovers Walking and Running." *Nature* 439 (January 5, 2006), pages 72–75.

Zihlman, Adrienne, and Douglas Cramer. "Human Locomotion." *Natural History*, vol. 85, no. 1 (January 1976), pages 64–69.

The Swarm

Evans, Howard Ensign. *Life on a Little-known Planet.* University of Chicago Press, 1984. Originally written in 1968, this is still the best non-technical book about insects I've seen. Chapter 8, "Paean to a Volant Voluptuary: The Fly," has some wonderful descriptions of swarms.

Oldroyd, Harold. *The Natural History of Flies.* Weidenfeld and Nicolson, 1964. This is a much more technical book than Evans's, but Chapter 20, "Swarms of Flies," is a good read.

The Evolution of the Teddy Bear

Gould, Stephen Jay. "Mickey Mouse Meets Konrad Lorenz." *Natural History*, vol. 88, no. 5 (May 1979), pages 30–36. This article was reprinted in Gould's *The Panda's Thumb* (McLeod, 1980).

Hinde, Robert, and L. A. Barden. "The Evolution of the Teddy Bear." *Animal Behaviour*, vol. 33, no. 4 (1985), pages 1371–73.

I'll Never Forget What's-His-Face

Bahrick, H. P., et al. "Fifty Years of Memory for Names and Faces: A Cross-Sectional Approach." *Journal of Experimental Psychology: General*, vol. 104, no. 1 (1975), pages 54–75. Every article about face memory mentions this classic high-school yearbook experiment.

Brownlee, Shannon. "What's in a Face." *Discover*, vol. 6 (February 1985), pages 39–51.

Loftus, Elizabeth. *Eyewitness Testimony.* Harvard University Press, 1979. If you're really serious about learning more about facial recognition, try Chapter 7 in this book.

Rubenstein, C. "Face." *Psychology Today*, vol. 17 (January 1983), pages 48–55.

An Uplifting Experience

Andersson, Malte, and Johan Wallander. "Kin Selection and Reciprocity in Flight Formation?" *Behavioral Ecology*, vol. 15 (January 2004), pages 158–62. Related birds might be more willing to lead the V.

Hainsworth, F. Reed. "Precision and Dynamics of Positioning by Canada Geese Flying in Formation." *Journal of Experimental Biology*, vol. 128 (1987), pages 445–62. This article focuses on the importance of the distance between wingtips rather than the angle of the V.

Lissaman, P. B. S., and Carl Shollenberger. "Formation Flight of Birds." *Science*, vol. 168 (1970), pages 1003–5. This is the article that started the recent flurry of interest in goose Vs.

When the Lavatory Becomes a Laboratory

Cahill, Spencer, et al. "Meanwhile Backstage." *Urban Life*, vol. 14, no. 1 (April 1985), pages 33–58. There is a frustrating shortage of easily accessible articles about bathroom behaviour, but this is a good read if you can find it. It describes the "interpersonal rituals and 'backstage' behaviours" of public-washroom occupants, including the well-known refusal of men at adjacent urinals to look at each other.

Gustavson, Andrew, Michael Dawson and Douglas Bonett. "Androstenol, a Putative Human Pheromone, Affects Human (*Homo sapiens*) Male Choice Performance." *Journal of Comparative Psychology*, vol. 101, no. 2 (1987), pages 210–12. The publication that showed that men apparently avoided lavatory stalls with androstenol-impregnated plastic squares on the doors.

Middlemist, R. D., E. S. Knowles and C. F. Matter. "Personal Space Invasions in the Lavatory: Suggestive Evidence for Arousal." *Journal of Personality and Social Psychology*, vol. 33, no. 5 (1976), pages 541–46. This is the notorious "periscope" paper.

Pedersen, D. M., S. Keithly and K. Brady. "Effects of an Observer on Conformity to Handwashing Norm." *Perceptual and Motor Skills*, vol. 62 (1986), pages 169–70. The hand-washing experiment.

Rusin, P., et al. "Reduction of Faecal Coliform, Coliform and Heterotrophic Plate Count Bacteria in the Household Kitchen and Bathroom by Disinfection with Hypochlorite Cleaners." *Journal of Applied Microbiology*, vol. 85 (1998), pages 819–28. This is the paper that establishes the non-lethality of the toilet seat.

Closet Science

Raloff, Janet. "The Electric Lifesaver Effect." *Science News*, vol. 134 (July 30, 1988), pages 78–79. Aside from short notes on the effect in books such as Jearl Walker's *Flying Circus of Physics*, this is the only article actually devoted to this phenomenon.

Letters from Heaven

There are many weather books with sections about snow and snow-storms, but two articles and one book should definitely be read:

Kepler, Johannes. *The Six-Cornered Snowflake*. Clarendon Press, 1966. This book, originally published in 1611, is a blend of science, philosophy and a hint of mysticism, but it does give you an insight into the mind of one of history's great scientists.

Knight, Charles, and Nancy Night. "Snow Crystals." *Scientific American*, vol. 228, no. 1 (January 1973), pages 100–107.

Knight, Nancy. "No Two Alike?" *Bulletin of the American Meteorological Society*, vol. 69, no. 5 (May 1988), page 496. If you can get your hands on this magazine, look at the photograph of the two "identical" snowflakes. It may be a piece of science history.

Feathers

Desmond, Adrian. *The Hot-Blooded Dinosaurs*. Warner Books, 1977. Chapter 6 features a good section on Archaeopteryx.

Feduccia, Alan, and Harrison Tordoff. "Feathers of Archaeopteryx: Asymmetric Vanes Indicate Aerodynamic Function." *Science*, vol. 203 (March 1979), pages 1021–22. A very short publication in which these two scientists point out the obvious: that the feathers of Archaeopteryx look as if they were suited to flight. Most of the remainder of this research is published in specialized journals.

Norell, Mark, and Xing Xu. "Feathered Dinosaurs." *Annual Review of Earth & Planetary Sciences*, vol. 33 (2005), pages 277–99.

This Chapter Is a Yawner

McCarthy, Paul. "Yawning to Breathe Free?" *Psychology Today* (February 1987), pages 9–10. This is a lay-language review of Provine's paper in *Ethology* below.

Platek, S., et al. "Contagious Yawning and the Brain." *Cognitive Brain Research*, vol. 23 (2005), pages 448–52. This is the paper that describes the parts of the brain activated during contagious yawning.

Provine, Robert. "Yawning as a Stereotyped Action Pattern and Releasing Stimulus." *Ethology*, vol. 72 (1986), pages 109–22. Provine is just about the only researcher who's actively investigating human yawning these days, and all his publications to date are in obscure scientific journals. This one provides the best review of yawning research over the years.

Provine, Robert. "Yawning." *American Scientist*, vol. 93 (2005), pages 532–39. This is the most recent review paper by Provine.

Just Walkin' in the Rain

I warn you—the following articles are for math buffs, but they are the only sources for this subject.

De Angelis, Alessandro. "Is It Really Worth Running in the Rain?" *European Journal of Physics*, vol. 8 (1987), pages 201–2. The message of this publication is summed up by the last sentence: "By running faster you get less wet, but the benefit ... does not justify the supplementary effort."

Deakin, Michael. "Walking in the Rain." *Mathematics Magazine*, vol. 45 (November–December 1972), pages 246–53. This paper establishes that you can get four times wetter than you should if you don't adopt the right strategies while walking, or running, through the rain.

Stern, S. A. "An Optimal Speed for Traversing a Constant Rain." *American Journal of Physics*, vol. 51, no. 9 (September 1983), pages 815–18. Even some physicists I've talked to have trouble getting through this paper.

Unsteady As She Goes

Allman, William F. "What Makes a Knuckleball Dance." In *Newton at the Bat*, edited by Eric Schrier and William Allman. Charles Scribner's Sons, 1984.

Robins, Benjamin. *New Principles of Gunnery*. London, 1742. This probably isn't in your branch library, but it is in the library of the National War Museum in Ottawa. The subtitle of the book is "An Investigation of the Difference in the Resisting Power of the Air to Swift and Slow Motions"—the origins of baseball science.

Watts, Robert, and Eric Sawyer. "Aerodynamics of a Knuckleball." *American Journal of Physics*, vol. 43, no. 11 (November 1975), pages 960–63. The first publication on the knuckleball.

Acknowledgments

In Victorian England, some of the most prominent scientists of the day took the time to give public lectures explaining the science of the everyday world. Not anymore: even those scientists today who appear on television and radio do not bother to spend much time on everyday things. Their science, the science of the twenty-first century, deals with forces and objects that are for the most part beyond our understanding.

We're a little poorer for being ignorant about the science of everyday life. For one thing, it's an approachable kind of science for anyone who's been intimidated by the subject since high school. (Quantum theory might be scary, but why we blink is not.) But more important, it makes life more interesting if you understand your world better. I guarantee that eating asparagus or yawning will never be the same once you've learned their scientific implications.

Every subject in this book is one that you're likely to encounter in the next few hours, days or perhaps months. The book will be a success if, after reading it, you look at these phenomena in a different way, or think about them a little longer than you normally would have.

I'm indebted to a number of researchers who took time from busy lives to talk to me about their work and to provide printed material, most of which would have been impossible to get any other way. These include Stanley Coren, Maurice Hershenson, Irvin Rock, Julia Chase, John Provine and John Stern.

I became convinced many years ago that science was much more interesting than most of the public knew, but I have to give credit to many others for showing me how to communicate that science in an engaging way. A partial list would include Joan Barberis, Sylvia Funston, Anita Gordon, Penny Park, Catherine Yolles, Meg Masters and Susan Folkins.

Wherever I end up after I die, I'm sure there will be a female editor standing next to me, telling me how to get it right. (I have no idea who of the above will be in which place.)

All this interest in science hasn't done my family much good. My father is still waiting for me to explain to him exactly how a computer chip works. My mother, who deserves the credit for making me into a birdwatcher, hasn't yet found out, at least from me, why black spots appeared on the surface of her whipped cream when she covered it overnight with aluminum foil. If nothing else, they have learned to be patient.